Interaction of Mechanics and Mathematics

Ingo Müller · Wolf Weiss

Entropy and Energy

A Universal Competition

With 76 Figures

Springer

Authors

Prof.Dr. Ingo Müller
Institut für Verfahrenstechnik
Technische Universität Berlin
Fasanenstraße 90
10623 Berlin
Germany

Dr. Wolf Weiss
Weierstraß-Institut für
Angewandte Analysis und Stochastik
Mohrenstr. 39
10117 Berlin
Germany

ISSN 1860-6245

ISBN-10   3-540-24281-3   Springer Berlin Heidelberg New York
ISBN-13   978-3-540-24281-9   Springer Berlin Heidelberg New York

Additional material to this book can be downloaded from http://extras.springer.com

Library of Congress Control Number: 2005925545

Springer is a part of Springer Science+Business Media

springeronline.com

© Springer-Verlag Berlin Heidelberg 2005

Typesetting: Data conversion by the authors.
Final processing by PTP-Berlin Protago-TeX-Production GmbH, Germany
Cover-Design: Erich Kirchner, Heidelberg
Printed on acid-free paper   89/3141Yu - 5 4 3 2 1 0

Preface

We perceive thermodynamic bodies as being subject to two − often opposing − tendencies: "energy and entropy" which may loosely be termed "cause and chance", or "determinacy and random walk". A high temperature speeds up the random walk and promotes the entropic tendency to grow, while a low temperature makes the body freeze in its energetic minima. We provide instructive examples from elementary thermodynamics and physico-chemistry for this competition. And we extrapolate the notion to non-standard thermodynamic subjects like shape memory, dissipation of the earth's atmosphere, and sociology.

We emphasize and revisit the works of the thermodynamic pioneers, in particular Clausius, Carathéodory, Boltzmann, Gibbs, and Planck. And we discuss the zeroth and third laws of thermodynamics and their limitations; also the pertinacious Gibbs paradox.

The accompanying CD provides instructive visualizations of the entropy-elastic spring and of phase transitions in fluids and shape memory alloys. The reader has numerous possibilities to influence the programs and thus develop an understanding for the thermodynamic principles involved in those phenomena.

The book is neither a textbook nor a monograph; it is something in-between. And while it may be read by beginners, it is probably best, if the reader has some prior knowledge of thermodynamics. For relief, and to atone for unavoidable technicalities, we have provided an account of the history of thermodynamics which is quite entertaining − at least that is what we think.

Some literature is quoted at the end of the book along with commentaries and annotations.

We thank Drs. A. Musolff and H. Sahota for their help with the figures and Dr. O. Kastner for letting us use his simulations of austenitic-martensitic transformations in shape memory alloys.

Mrs. M. Hentschel's cheerful cooperation in typing − and retyping − the manuscript is grateful acknowledged.

Berlin, September 10, 2004   *Ingo Müller and Wolf Weiss*

Contents

1

Origin of entropy in the work of Clausius

Considering the fundamental importance which we now attribute to entropy in physics and chemistry and natural philosophy one may find it odd that Clausius discovered the notion in the context of heat engines. We must realize, however, that thermodynamics started out as the science of the "motive power of heat" and the very name of the science − thermodynamics − recalls that origin.

Clausius' proof of the existence of entropy and his investigation of its properties are very much "a physicist's argument" starting from a loosely worded axiom. As such, the proof is looked upon with disdain by analysts. Even so, in this chapter we retrace the steps of Clausius. After all, he was first, and physicists at least find his arguments convincing; they love them!

1.1 Historical background

In the 1820's thousands of steam engines were already installed, most of them in England and most of them were used for pumping water out of coal mines. Thus coal could be brought up from a greater depth − previously inaccessible − and at first it did not matter much that a little part of that coal was used up by the boiler of the engine. Increasingly, however, steam engines were driving machines like lathes, drills, spinning wheels, looms, etc. Thus coal became a commodity to be paid for by the owners of the machines. Therefore, naturally, the desire arose to increase the efficiency of the steam engines so that more work could be gained from a given amount of coal.

Everything seemed possible: After all, the steam engine had been developed by blacksmiths and amateurs − and a good job they did (!) − and what had emerged was an engine that worked with liquid water and water vapour and was heated and cooled at constant pressures. Perhaps it would be better for efficiency to use mercury, or even air? And perhaps it might be better to apply and withdraw the heat at constant volumes, or at constant temperatures?

Those were questions for thermodynamicists except that the science of thermodynamics did not exist yet and that there were no thermodynamicists. Carnot made a heroic premature attempt and despite the fact that he made wrong assumptions and wrong derivations from those assumptions[1] he was able to come up with a correct result. We may now express that result in the words:

The efficiency of a Carnot engine is maximal and universal.

A Carnot engine exchanges heat at two temperatures only and no other engine – working in the same range of temperature – has a bigger efficiency than a Carnot engine. Moreover, that maximal efficiency is universal, i.e. independent of the working substance, be it water, air, or steel.

Carnot believed in the caloric theory of heat that prevailed at his time. By that theory the heat passing through the engine from boiler to cooler is unchanged in amount. However, that assumption is quite wrong as we shall see. In fact it is something of a miracle that Carnot, despite his erroneous reasoning, could come up with his correct result about the Carnot efficiency.

1.2 Thermodynamic background

Carnot's difficulty was that he did not know the balance of energy in a form including heat, the so-called *first law of thermodynamics*. That law was discovered only in or around 1840 by Mayer.[2] In a modern form – and applied to a closed body[3] – it reads

$$\frac{\mathrm{d}(U + K)}{\mathrm{d}t} = \dot{Q} + \dot{A} \quad \bigg| \quad 1^{\text{st}} \text{ law of thermodynamics.} \qquad (1.1)$$

The 1st law states that the rate of change of energy of the body – internal energy U and kinetic energy K of the flow field – is equal to the sum of heating and working. Typically $\dot{Q}$ and $\dot{A}$ represent the rate of exchange of internal energy through the boundary and the power of stresses acting on the boundary of the body, respectively. [$\dot{Q}$ may also contain the rate of absorption of radiative energy inside the body and $\dot{A}$ may contain the power of gravitational, inertial, or electro-magnetic forces inside the body. However, these phenomena did not enter into Clausius's arguments.]

---

[1] This iconoclastic claim is documented in Chap. 21 where we give a brief history of thermodynamics concentrating on the idiosyncrasies of the pioneers and the difficult birth of the thermodynamic concepts.

[2] See Chap. 21 for a short history of thermodynamics.

[3] In thermodynamics a closed body has a boundary that is impenetrable for mass so that its mass is constant.

A cycle is a process of heating and working between times t_i and t_f that begins and ends with the same values of $U + K$, where K is usually supposed to be zero at the beginning and the end. Therefore we have

$$Q_\bigcirc = -A_\bigcirc, \qquad (1.2)$$

where $Q_\bigcirc = \int_{t_i}^{t_f} \dot{Q} \, dt$ and $A_\bigcirc = \int_{t_i}^{t_f} \dot{A} \, dt.$[4] In a heat engine, where work is produced, $A_\bigcirc$ is negative, while in a refrigerator, where work is consumed, $A_\bigcirc$ is positive. In both cases $Q_\bigcirc$ consists of a positive part Q_+ and a negative part Q_-, the former is the heat $\int \dot{Q} \, dt$ consumed by the boiler and the latter is released by the cooler. In a heat engine the efficiency is defined as

$$e = \frac{|A_\bigcirc|}{Q_+}. \qquad (1.3)$$

A possible realization of a heat engine is shown in Fig. 1.1$_{left}$. In that case the "closed body" to which the first law is applied, is a fixed amount of gas which is pumped round and round in a cycle of compression, heating, decompression and cooling as indicated. Compression and decompression in that engine are performed adiabatically, while heating and cooling occur isobarically. The working $A_\bigcirc$ may be taken off from the crankshaft.

In general working and heating on the boundary of a body will create havoc inside the body: A turbulent flow field, and strongly inhomogeneous fields of mass density, pressure and temperature. We may say that the internal equilibrium of the body is disturbed. But, if the working and heating are applied carefully and slowly, there will be no appreciable flow field in the interior, and pressure and temperature will be essentially homogeneous while slowly changing in time; the internal equilibrium will thus prevail during the slow application of working and heating. Such slow processes may therefore be called *equilibrium processes*; they are also called *reversible processes* − for a reason to be explained − or *quasistatic processes*. The opposites are non-equilibrium processes, irreversible processes and rapid processes.

How slow is slow? Well, that depends! As rough rules we may say that
- slow means slow compared to the propagation of sound,
- the fields, like pressure and temperature, do not change appreciably during the body's relaxation time, like the mean time of free flight of an atom in a gas,
- gradients of the fields are small on characteristic lengths of the body, like the mean free path in a gas.

In this manner even the movement of a piston in the engine of a racing car may be considered as slow enough to be treated as reversible in the manner we proceed to describe. After all, the piston is still a lot slower than sound.

---

[4] Q and A without dots, if they occur, are time integrals over heating $\dot{Q}$ and working $\dot{A}$ and we speak of heat transmitted to the body and of work done to it in an interval of time.

If the working is applied *slowly* against the pressure p of the gas by a change of volume V we have

$$\dot{A} = -p\frac{dV}{dt}. \qquad (1.4)$$

In that special case the kinetic energy can be neglected and the pressure field and the temperature field inside the body may be considered as homogeneous. Therefore the first law (1.1) reduces to the form

$$\frac{dU}{dt} = \dot{Q} - p\frac{dV}{dt}. \qquad (1.5)$$

In a slow process the internal energy $U(t)$ and the pressure $p(t)$ depend on time, because they are functions of the volume $V(t)$ and temperature $T(t)$, different functions for different materials.[5] The relations

$$U = U(T,V) \quad \text{and} \quad p = p(T,V) \qquad (1.6)$$

are called caloric and thermal equations of state respectively and, for a given material, they must be determined experimentally. There is only one class of materials for which these functions are known explicitly and analytically and that is for ideal gases in which case we have[6]

$$U = mz\frac{k}{\mu}(T - T_R) + U(T_R) \quad \text{and} \quad p = \frac{m}{V}\frac{k}{\mu}T. \qquad (1.7)$$

m is the mass of the gas and μ is the mass of its molecules. T is the absolute temperature, nowadays measured in K for Kelvin. k is the Boltzmann constant $k = 1.38 \cdot 10^{-23}\frac{J}{K}$, and z has values $\frac{3}{2}$, $\frac{5}{2}$ and 3 for one-, two-, or more-atomic gases, respectively. T_R is an arbitrary reference temperature, usually chosen as 25°C or 298.15 K, at least by chemists. Note that U is independent of V in an ideal gas and depends only on T.

A slow process $V(t)$, $T(t)$ is called *reversible* because – according to (1.4) through (1.6) – the heating is reversed (sic) when the process is reversed, i.e. when $\frac{dV}{dt}$ and $\frac{dT}{dt}$ change signs. Such a reversible process may be represented by a line in a (V,T)-diagram, or a (p,V)-diagram, since only two functions of time characterize the process. In particular, if a Carnot engine runs reversibly, the cycle may be represented by the graph of Fig. 1.1$_{\text{right}}$ consisting of the branches 12 and 34 of the isotherms $T_H = const$ and $T_L = const.$, and of the branches 23 and 41 of two adiabates, where no heating occurs.

---

[5] In a rapid and strongly non-homogeneous process $U(t)$ and $p(t)$ might depend on the *histories* of the mass density ρ and of temperature T in the points $\mathbf{x}\epsilon V$, and on the gradients $\frac{\partial\rho}{\partial x_i}$ and $\frac{\partial T}{\partial x_i}$ in those points.

[6] Actually that knowledge also comes from experiments. We learn about them in high school; the experiments were done by Boyle, Mariotte, Gay-Lussac and Joule and Thomson centuries ago.

For an ideal gas as the working substance we may easily calculate the contributions to $A_\circlearrowright$ and $Q_\circlearrowright$ in a reversible Carnot cycle from (1.4) through (1.7)

$$A_{12} = -m\tfrac{k}{\mu}T_H \ln \tfrac{V_2}{V_1} \qquad Q_{12} = m\tfrac{k}{\mu}T_H \ln \tfrac{V_2}{V_1} > 0$$

$$A_{23} = mz\tfrac{k}{\mu}(T_L - T_H) \quad Q_{23} = 0$$

$$A_{34} = -m\tfrac{k}{\mu}T_L \ln \tfrac{V_4}{V_3} \qquad Q_{34} = m\tfrac{k}{\mu}T_L \ln \tfrac{V_4}{V_3} < 0 \tag{1.8}$$

$$A_{41} = mz\tfrac{k}{\mu}(T_H - T_L) \quad Q_{41} = 0 \ .$$

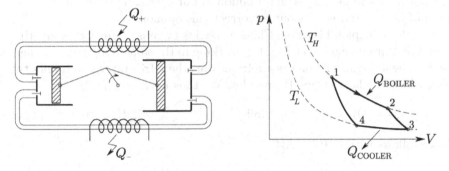

Fig.1.1 Left: The prototypical heat engine with adiabatic compression and decompression and isobaric heat exchange.
(Not a Carnot engine!)
Right: The graph of a reversible Carnot cycle in a (p, V)-diagram.

Hence we obtain the efficiency of a slow Carnot cycle with an ideal gas, again by use of (1.4) through (1.7)

$$e = \frac{m\tfrac{k}{\mu}T_H \ln \tfrac{V_2}{V_1} + m\tfrac{k}{\mu}T_L \ln \tfrac{V_4}{V_3}}{m\tfrac{k}{\mu}T_H \ln \tfrac{V_2}{V_1}} = 1 - \frac{T_L}{T_H}. \tag{1.9}$$

We confirm that e depends only on the two temperatures. In particular it is independent of the type of ideal gas. But we do not know yet whether the same formula holds for vapours, liquids or solids, irrespective of the form of their caloric and thermal equations of state (1.6). Actually, seeing that in the derivation of the Carnot efficiency (1.9), we have made extensive use of the ideal gas laws (1.7), it might seem that $e = 1 - \frac{T_L}{T_H}$ is restricted to ideal gases. However, that is not so as Clausius was able to prove.

1.3 The second law of thermodynamics

Universal efficiency of a Carnot engine

The concept of entropy was discovered by Clausius in the early 1850's, and it was also he who invented the name. At first Clausius was primarily concerned with heat engines and thermodynamic cycles. Like Carnot he asked which cycle might be best suited for converting heat into work and which substance should be used for an optimal conversion. Clausius bases his argument on the plausible axiom

Heat cannot pass by itself from a cold to a hot body,

which has become known as the second law of thermodynamics.

Clausius knew Carnot's work and he is quite polite about it; he says mildly: "Carnot ... has formed a peculiar opinion about cause and effect in the transition of heat." And he sets out to correct this opinion.

In order to exploit his axiom Clausius begins by considering two competing reversible Carnot engines, cf. Fig. 1.2, working in the same temperature range, one as a heat engine and one as a refrigerator; the refrigerator I consumes the work which the heat engine II provides. We thus have from (1.2)

$$-A_{\bigcirc}^I = Q_B^I - |Q_C^I| < 0 \quad \text{and} \quad -A_{\bigcirc}^{II} = Q_B^{II} - |Q_C^{II}| > 0. \qquad (1.10)$$

Hence follows with $-A_{\bigcirc}^I = A_{\bigcirc}^{II}$

$$Q_B^I - |Q_C^{II}| = |Q_C^I| - Q_B^{II}. \qquad (1.11)$$

We assume that $Q_B^I - |Q_C^{II}| > 0$ holds, so that the refrigerator I accepts more heat at T_L than the heat engine II gives off there. By (1.11) this also means that at T_H the refrigerator I gives off more heat than the heat engine II accepts there. The net result — after both engines have had an equal number of revolutions — is a transition of heat from the low temperature T_L to the high temperature T_H, and that is forbidden by the second law. Therefore $Q_B^I - |Q_C^{II}|$ cannot be positive. But the alternative $Q_B^I - |Q_C^{II}| < 0$ is also impossible. To be sure, when engine I absorbs less heat at T_L than engine II provides there, the two engines effect a transition from warm to cold which is *not* forbidden. However the engines are reversible and a reversal will bring us back to the previous impossible situation.

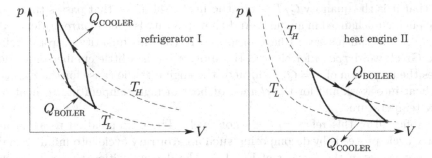

Fig.1.2 Clausius' competing reversible Carnot engines.

Therefore we must conclude that

$$Q_B^I = |Q_C^{II}| \quad \text{hence} \quad |Q_C^I| = Q_B^{II} \tag{1.12}$$

and it follows that we have

$$\frac{A_\circlearrowleft^I}{|Q_C^I|} = \frac{-A_\circlearrowleft^{II}}{Q_B^{II}}, \quad \text{hence by (1.3)} \quad e^I = e^{II}. \tag{1.13}$$

Since both engines may work with different substances, we conclude from $(1.13)_2$ that the efficiency of a reversible Carnot engine is *universal*, i.e. independent of material. If, in particular, the material is an ideal gas, we know that e is given by (1.9). Therefore it is *always* given by (1.9) and we have

$$e = 1 - \frac{T_L}{T_H} \tag{1.14}$$

irrespective of material. Thus Clausius proved what Carnot had asserted.

Entropy

However, Clausius went far beyond this point. He proceeded to discover entropy and then liberate this new concept from its origin in cycles and, in particular, Carnot cycles.

With $-A_\circlearrowleft = Q_B - |Q_C|$ the efficiency of a heat engine may be written in the form

$$e = 1 - \frac{|Q_C|}{Q_B}. \tag{1.15}$$

Comparing (1.14) and (1.15) we conclude

$$\left.\frac{Q}{T}\right|_{\text{Boiler}} = \left.\frac{|Q|}{T}\right|_{\text{Cooler}} \quad \text{or} \quad \left.\frac{Q}{T}\right|_{\text{Boiler}} + \left.\frac{Q}{T}\right|_{\text{Cooler}} = 0. \tag{1.16}$$

so that it is the quantity Q/T – not the heat itself (!) – that passes through the engine unchanged in amount. And that quantity is called *entropy*, denoted by $S \equiv \frac{Q}{T}$. Clausius says: I have proposed to call this quantity entropy from the Greek word $\tau\rho o\pi\acute{\eta}$ for change. His motivation is a little cumbersome. He sees the transition of $S = Q/T$ through the engine as the cause for the *change* of heat into work and for the *change* of heat of high temperature to heat of low temperature.

Sofar all of this refers to a Carnot cycle. The extrapolation to an arbitrary cycle proceeds by decomposing such an arbitrary cycle into infinitesimal Carnot cycles in the manner of Fig. 1.3. The decomposition has to be such that the area of each strip equals the area of the smooth cycle which it replaces and, moreover, such that each single zig-zag has equal areas above and below the smooth curve. In that manner it is guaranteed that the smooth cycle and the zig-zag cycle generate the same work and that each line element of the smooth cycle requires the same heat. The idea is clear: Each strip satisfies $(1.16)_2$ so that the integral of $\frac{\dot{Q}dt}{T}$ from the beginning of the cycle at time t_i to its end at time t_f gives zero

$$\int_{t_i}^{t_f} \frac{\dot{Q}dt}{T} = 0. \tag{1.17}$$

And just as we used to have $S = \frac{Q}{T}$ on the boiler or cooler side of the finite Carnot cycles, we now have for one of the infinitesimal Carnot cycles

$$dS = \frac{\dot{Q}dt}{T} \qquad \text{or} \qquad \frac{dS}{dt} = \frac{\dot{Q}}{T}. \tag{1.18}$$

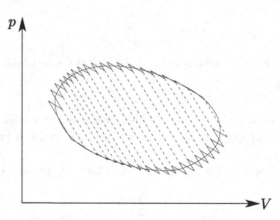

Fig. 1.3 Smooth cycle decomposed into infinitesimal Carnot cycles.

And this holds for the corresponding element of the smooth cycle as well, since its $\dot{Q}dt$ and T are equal to those of the Carnot cycle. Elimination of $\dot{Q}dt$ between (1.18) and the first law (1.5), appropriate to a slow process, gives

$$dS = \frac{1}{T}\left(dU + pdV\right), \tag{1.19}$$

so that we may say that $1/T$ is an integrating factor of the differential form $dU + pdV$; the corresponding potential is the entropy S, a state function, depending on U and V, or V and T in a materially dependent manner. Equation (1.19) is called the *Gibbs equation* even though it appears already in Clausius' work. Gibbs extrapolated this equation to mixtures and fully exploited that case.

With U and p given by the caloric and thermal equations of state (1.7) of an ideal gas we may calculate $S(T, V)$ for an ideal gas by integration of the Gibbs equation. We obtain

$$S = m\left[\frac{k}{\mu}\ln\left(T^z V\right) + const\right]. \tag{1.20}$$

Maximum efficiency

In order to show that a slow Carnot cycle has the biggest efficiency among all cycles with the same temperature range $T_L \leq T \leq T_H$, we choose a (T, S)-diagram for representing the graph of a cycle. By (1.18) the boiler-heat Q_+ equals the area below the upper part C_U of the graph, while the cooler-heat $|Q_-|$ equals the area under the lower part C_S of the graph, cf. Fig. 1.4$_{left}$. We proceed in two steps:

i)   Given Q_+, the efficiency

$$e = 1 - \frac{|Q_-|}{Q_+}$$

is maximal when $Q_- = \int_{C_S} TdS$ is minimal which is obviously the case when all the heat comes out at $T = T_L$, see Fig. 1.4$_{center}$, so that $Q_- = T_L\Delta S$.

ii)   If we wish to make Q_- smaller yet, we must decrease ΔS. But this must be done in such a manner that the area below C_U remains unchanged. Rather obviously this requires that we deform that area into a rectangle, cf. Fig. 1.4$_{right}$. In this manner we have arrived at a cycle consisting of two isotherms and two isentropes, i.e. by (1.18), two adiabates. This is therefore the graph of a Carnot cycle.

We do feel that this proof — in two steps — of maximal efficiency for a slow Carnot process may not satisfy a mathematician. It seems that an argument like this should be couched in the language of variational calculus.

The above argument is our own. Clausius proves the result by recourse to the second law. He shows that maximum efficiency for a Carnot cycle is necessary lest heat pass from cold to hot by itself.

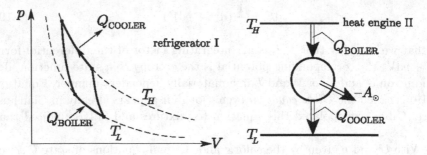

Fig.1.4 On the maximum efficiency of a slow Carnot cycle.

Growth of entropy

Now we rejoin Clausius' argument and proceed to investigate competing Carnot cycles like those in Fig. 1.2 but such that one of them, the heat engine, is not reversible, so that the heating is not reversed in sign when the cycle is reversed. In that case we cannot draw a graph for the heat engine cycle in a (p, V)-diagram. Instead we represent the situation schematically as in Fig. 1.5.

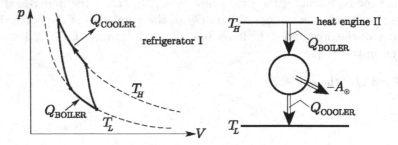

Fig.1.5 Two competing Carnot engines.
Left: The reversible refrigerator I
Right: The irreversible heat engine II.

We can still conclude that $Q_B^I - |Q_C^{II}| > 0$ must be excluded just like before when the engines were considered reversible. But now we must admit the alternative $Q_B^I - |Q_C^{II}| < 0$, since this case cannot now be rendered contradictory to the second law by reversing the engines. After all, the heat engine *is* not reversible. Therefore we have to assume

$$Q_B^I < |Q_C^{II}| \qquad \text{hence} \qquad |Q_C^I| < Q_B^{II}$$

and we conclude that we must have

$$\frac{A_\circlearrowleft^I}{|Q_C^I|} > \frac{A_\circlearrowleft^{II}}{Q_B^{II}}, \quad \text{hence by (1.3)} \quad e^I > e^{II}. \tag{1.21}$$

It follows that the efficiency of the irreversible engine II is smaller than the one of the reversible engine. The latter, of course, is still equal to $1 - \frac{T_L}{T_H}$. Therefore $(1.21)_1$ and $(1.10)_2$ imply

$$\left.\frac{Q}{T_0}\right|_{\text{Boiler}} + \left.\frac{Q}{T_0}\right|_{\text{Cooler}} < 0,$$

whence follows, much like before[7]

$$\int_{t_i}^{t_f} \frac{\dot{Q}dt}{T_0} < 0, \tag{1.22}$$

which replaces (1.17). This inequality holds for any arbitrary irreversible cycle.

Let us consider a cycle in $t_i \leq t \leq t_f$ which is slow and reversible during the partial interval $t_i \leq t \leq t_m$ but fast and irreversible during the rest of the time. In that case (1.22) is best decomposed in the form

$$\int_{t_i}^{t_m} \frac{\dot{Q}dt}{T} + \int_{t_m}^{t_f} \frac{\dot{Q}dt}{T_0} < 0,$$

because, by (1.18), the first integral can be integrated to give $S_m - S_i$ or $S_m - S_f$ since, after all, we deal with a cycle so that $S_f = S_i$ holds. Therefore we obtain

$$S_f - S_m > \int_{t_m}^{t_f} \frac{\dot{Q}dt}{T_0}. \tag{1.23}$$

We may express this by saying that between the beginning and the end of an irreversible process the entropy change is bigger than $\int \frac{\dot{Q}dt}{T_0}$. If the irreversible process is infinitesimally short we obtain

$$\frac{dS}{dt} > \frac{\dot{Q}}{T_0} \quad \text{or combining} \quad \boxed{\frac{dS}{dt} \geq \frac{\dot{Q}}{T_0} \quad \begin{array}{l} 2^{\text{nd}} \text{ law of} \\ \text{thermodynamics} \end{array}}, \tag{1.24}$$

this with (1.18)

where the equality holds when the process is reversible; in that case T_0 may be replaced by T, the homogeneous temperature throughout the body.

The frame in (1.24) contains the mathematical expression for the second law, at least as far as the Clausius argument goes.

---

[7] Recall that $\dot{Q}$ is the heating at the boundary of the body which runs through the cycle and that the temperature is the temperature of the boundary. In a reversible, i.e. slow process the temperature is homogeneous throughout the body; but for an irreversible process we must distinguish T in the interior — possibly a wildly changing and strongly non-homogeneous field — and T on the boundary which we denote by T_0. In all of Clausius' arguments T_0 is homogeneous on the boundary.

In particular, for an adiabatic process, where $\dot{Q}$ in (1.24) vanishes, the entropy cannot decrease.

A suggestive interpretation of the inequality

During a rapid process the interior of the body may contain kinetic energy in its flow field — usually turbulent. That flow is slowed down by friction, and its kinetic energy is thereby converted into internal energy which leads to a rise of temperature. Thus the decrease of kinetic energy is related to heating or, better, *internal* heating. The internal heating at the local temperature increases the entropy beyond the change $\dot{Q}/T_0$ which occurs in (1.24) and which is due to heating through the surface at the surface temperature T_0.

Friction thus contributes to the inequality (1.24) but that is not all. Indeed, in a rapid process the temperature field may be strongly non-homogeneous inside the body. The temperature differences are decreased by an exchange of internal energy between warm parts and cold parts; from warm to cold. Such internal cooling of warm parts and heating of cold parts increases the entropy of the parts involved. Thus $\frac{dS}{dt}$ becomes bigger than the $\dot{Q}/T_0$ which, after all, is due to heat conduction through the surface only.

This may suffice for a first interpretation of the inequality (1.24). Later, in Chaps. 3 and 17, we shall be more specific and we shall then also learn more about the "mechanisms" of entropy production, viz. internal friction and internal heat conduction.

Extrapolation of the 2nd law

Recognizing that the entropy must grow in an irreversible adiabatic process Clausius has revealed a teleological tendency of nature. This was much discussed at his time and is still much discussed today — and not only by physicists and mathematicians! Philosophers, sociologists and historians have let themselves be fascinated by entropy and its growth property. They extrapolate the entropic tendency to grow to their fields of knowledge and not all of their arguments make good sense to a natural scientist. In Chap. 21, describing the history of thermodynamics, we shall give an account of some of the more outrageous and colorful notions.

Despite this disparaging comment we firmly believe that thermodynamic ideas — in particular those of Clausius — may find an application outside thermodynamics, even outside physics, namely in the fields of economy and sociology. Too obvious are the irreversible tendencies occuring in those fields to be ignored. Chap. 20 presents an attempt at the formulation of *sociothermodynamics*, a theory that may conceivably be able to describe the phenomena of segregation and desegregation in an ethnically mixed population.

2

Entropy in the work of Carathéodory. Absolute temperature

Carathéodory felt uncomfortable with the notions of heat and heating and therefore Clausius' axiom, discussed in Sect. 1.3, meant nothing to him. He had to find an alternative axiom and an alternative route to entropy. Being an analyst he found that route in his research on differential forms and their integrating factors. Altogether Carathéodory's arguments are more formally mathematical than those of Clausius, and therefore his analysis highlights some concepts of thermodynamics which Clausius took for granted. Therein lies the lasting value of Carathéodory's work. In this chapter we review his ideas and − for definiteness − we illustrate them for visco-elastic bodies.

2.1 Empirical temperature

Among the concepts discussed clearly and properly by Carathéodory is the notion of temperature; empirical and absolute temperature. Temperature is defined as *the* quantity which is equal in two bodies in thermal contact. And since two bodies in thermal contact appear to be equally hot − after they have reached equilibrium − we may say that temperature measures how hot a body is.[1] Any such measure will do. We denote one of them by ϑ and call it the empirical temperature; all we require is that there is a one-to-one correspondence between ϑ and hotness such that ϑ is bigger, when the body is hotter.[2]

In terms of ϑ we may write the thermal and caloric equations of state of an ideal gas of mass m in the form

$$p = \frac{m}{V}\frac{1}{\mu}f(\vartheta) \quad \text{and} \quad U = mz\frac{1}{\mu}f(\vartheta) + const. \tag{2.1}$$

---

[1] The concept of temperature and its measurement did not come easy to the scientists in the 17th century who introduced it. See Chap. 21, where we give a brief history of temperature within the history of thermodynamics.

[2] It is also possible to let ϑ decrease with increasing hotness. Such was the case in the original Celsius scale, cf. Chap. 21.

These relations have been found experimentally in the 17th through 19th century. In particular, it turned out that $f(\vartheta)$ is the same function for all ideal gases and that it is monotonically increasing. Note that the values of that function may be determined — and were determined — for all values of the empirical temperature ϑ from measurements of p and V.

z in (2.1) is either $\frac{3}{2}$, $\frac{5}{2}$, or 3 depending on whether the gas has one-, two-, or more-atomic molecules. μ is the mass of the molecules.

2.2 First law for visco-elastic bodies

Feeling uncomfortable with heat and heating Carathéodory confined his attention to adiabatically isolated bodies. Thus the first law (1.1) reduces to

$$\frac{\mathrm{d}(U+K)}{\mathrm{d}t} = \dot{A}. \qquad (2.2)$$

We simplify our task by neglecting the kinetic energy and by assuming the processes inside the body to be homogeneous. Also we restrict the attention to thermo-visco-elastic bodies. [This means misrepresenting Carathéodory somewhat, because his scope is universal. But we gain in clarity what we lose in generality; we hope.]

Let the working be done by external pressures $p_\alpha (\alpha = 1, 2...n)$ — components of the pressure tensor — on the surface and let those forces effect changes of the strains $V_\alpha (\alpha = 1, 2...n)$. Thus the working may be written as

$$\dot{A} = - \sum_{\alpha=1}^{n} p_\alpha \frac{\mathrm{d}V_\alpha}{\mathrm{d}t}. \qquad (2.3)$$

We assume that in general the pressures are applied
- slowly enough to maintain homogeneous conditions inside the body and to prevent flows with an appreciable kinetic energy, but
- fast on the time scale of the relaxation of pressures and internal energy of the body.

In other words we assume that the external pressures during the processes are equilibrated by the visco-elastic pressures of the body. However, there is still creeping motion and temperature change due to relaxation processes. Such is the case in homogeneous visco-elastic bodies which are given to processes like creep or stress-relaxation.

Under those conditions the visco-elastic equations of state imply that U and p_α at time t depend on the histories $\vartheta(t-s)$, $V_\alpha(t-s)$ of temperature and strains. $t-s$ is any time in the past as far back as the body remembers, i.e. as far back as the largest relaxation time in the body. We write

$$U(t) = \mathfrak{U}\left[\vartheta(t-s),\, V_\beta(t-s)\right],$$

$$p_\alpha(t) = \mathfrak{p}_\alpha\left[\vartheta(t-s),\, V_\beta(t-s)\right],$$

(2.4)

where the script letters denote functionals. The first law reads

$$\frac{d}{dt}\mathfrak{U}\left[\vartheta(t-s),\, V_\beta(t-s)\right] = -\sum_{\alpha=1}^{n}\mathfrak{p}_\alpha\left[\vartheta(t-s),\, V_\alpha(t-s)\right]\frac{dV_\alpha}{dt}.$$

(2.5)

One observation is obvious: If this equation holds for a process $\vartheta(t)$, $V_\alpha(t)$, it does not hold for the reversed process in which $\frac{d\vartheta}{dt}$ and $\frac{dV_\alpha}{dt}$ are changed in sign, because the histories differ for the forward and backward process. This means that processes in visco-elastic bodies are *irreversible* in general.

There is an exception and that exception occurs for quasistatic processes, i.e. processes which are so slow that a change of $\vartheta(t)$, $V_\alpha(t)$ during the reach of memory of the body is negligible. In that case the caloric and thermal state *functionals* reduce to state *functions* and (2.4) is simplified to

$$U(t) = U(\vartheta(t),\, V_\beta(t))$$

$$p_\alpha(t) = p_\alpha(\vartheta(t),\, V_\beta(t))$$

(2.6)

Under those circumstances the first law reads

$$\frac{dU\,(\vartheta, V_\beta)}{dt} = -\sum_{\alpha=1}^{n} p_\alpha\,(T, V_\beta)\,\frac{dV_\alpha}{dt},$$

(2.7)

and now, obviously, this equation is satisfied for the forward and the backward process, i.e. the quasistatic process is *reversible*.

Concerning $(2.6)_1$ Carathéodory assumes that by the caloric equation of state U increases monotonically with ϑ, so that we have

$$\left(\frac{\partial U}{\partial \vartheta}\right)_{V_\alpha} > 0.$$

(2.8)

2.3 Carathéodory's axiom

We continue to consider homogeneous visco-elastic bodies and formulate Carathéodory's inaccessability axiom − his version of the second law − in a form appropriate to that special case:

In the neighbourhood of an arbitrary initial state ϑ^i, V_α^i there are states, which cannot be reached by an adiabatic process $\vartheta(t)$, $V_\alpha(t)$.

At first sight this axiom seems to be a pretty formal condition on the functions $\vartheta(t), V_\alpha(t)$ that satisfy the functional equation (2.5) And yet the axiom is eminently plausible. Let us consider:

For illustration we consider a process which begins quasistatically with ϑ^i, V_α^i and ends quasistatically with ϑ^f, V_α^i, so that the final strains are identical to the initial ones. In-between the strains may have reached arbitrary values and the process may have been strongly irreversible. What can we say about the final temperature ϑ^f?

The final temperature ϑ^f follows by integration of the 1st law (2.5) between t^i and t^f over a process that begins and ends quasistatically

$$U\left(\vartheta^f, V_\alpha^i\right) - U\left(\vartheta^i, V_\alpha^i\right) = -\sum_{\alpha=1}^{n} \int_{t^i}^{t^f} \mathfrak{p}_\alpha\left[(\vartheta(t-s), V_\beta(t-s)\right] \frac{dV_\alpha}{dt} dt. \quad (2.9)$$

Thus ϑ^f depends on the working during the process. The only reasonable assumption seems to be that, because of viscosity and friction, we cannot hope to cool the body by working. This assumption expresses the old adage: "Friction heats". Therefore the states ϑ^f, V_α^i with $\vartheta^f < \vartheta^i$ are inaccessible from ϑ^i, V_α^i in an adiabatic process. Note that, since ϑ does not decrease in the processes considered in (2.9), neither does U, because of (2.8).

Other illustrations of the sensibleness of Carathéodory's axiom can be constructed, while no plausible contradiction has ever been found. Therefore the inaccessibility axiom is accepted as expressing the essence − under adiabatic conditions − of ubiquitous dissipation.

2.4 Exploitation of the axiom

Carathéodory's axiom holds in particular for reversible processes, where the first law (2.5) may be written as a differential equation for $U(t)$ and $V_\alpha(t)$ in the form[3]

$$\frac{dU}{dt} + \sum_{\alpha=1}^{n} p_\alpha\left(U, V_\beta\right) \frac{dV_\alpha}{dt} = 0. \quad (2.10)$$

Now, differential equations of that type was something that Carathéodory knew very well. He had a lemma − Carathéodory's *lemma on Pfaffian forms* − which reads:

---

[3] Here and in the sequel we sometimes switch between ϑ as a variable and U. This is possible because of the assumption (2.8). Later we shall assume, without much comment, that all p_α's depend at least on *one* strain V_β. Carathéodory is very careful to mention all such assumptions in order to exclude degenerate cases, but we are less strict for brevity. If degenerate cases occur in this book they will be treated separately.

Consider the differential equation $\frac{dx_0}{dt} + \sum\limits_{\alpha=1}^{n} X_\alpha \frac{dx_\alpha}{dt} = 0$ *with* $X_\alpha = X_\alpha(x_0, x_1...x_n)$. *If there exist points in the neighbourhood of an arbitrary point* $x_0, x_1...x_n$, *which cannot be connected to that point by a solution* $x_0(t), x_1(t)...x_n(t)$ *of the differential equation, then the Pfaffian form* $dx_0 + \sum\limits_{\alpha=1}^{n} X_\alpha dx_\alpha$ *has integrating factors* $\Lambda(x_0, x_1...x_n)$.

It is clear that the lemma is relevant to thermodynamics, if we accept Carathéodory's axiom. Indeed, it implies that $dU + \sum\limits_{\alpha=1}^{n} p_\alpha dV_\alpha$ has integrating factors $\Lambda(U, V_\beta)$ so that there exists a "potential" $S_\Lambda(U, V_\beta)$:

$$dS_\Lambda = \Lambda \left\{ dU + \sum_{\alpha=1}^{n} p_\alpha dV_\alpha \right\}. \tag{2.11}$$

We do not repeat the proof of Carathéodory's lemma. It can be found in the mathematical literature or, most appropriately, in Carathéodory's work.[4] However, we wish to remind the reader of some features concerning integrating factors which are important for our argument: First of all, it is true that, if a Pfaffian form has one integrating factor, it has many; their ratio is an arbitrary function of S_Λ. Secondly, a Pfaffian form in only *two* variables, i.e. with $n = 1$, always has integrating factors; no inaccessibility axiom is needed to establish their existence in that case.

After the existence of the integrating factors of $dU + \sum\limits_{\alpha=1}^{n} p_\alpha dV_\alpha$ is thus established, Carathéodory proceeds to show that one of them is a universal function of temperature.

2.5 One integrating factor is a universal function of ϑ. Entropy

We consider two visco-elastic bodies I and II of the type considered heretofore, i.e. with homogeneous fields of temperature and strains. Thus (2.11) holds for both of them and we have

$$dS_\Lambda^I = \Lambda^I \left(dU^I + \sum_{\alpha=1}^{n} p_\alpha^I dV_\alpha^I \right) \quad \text{and} \quad dS_\Lambda^{II} = \Lambda^{II} \left(dU^{II} + \sum_{\alpha=1}^{m} p_\alpha^{II} dV_\alpha^{II} \right).$$
$$\tag{2.12}$$

The number of strain components − n and m respectively − may be different.

---

[4] References may be found at the end of this book.

Also we consider the compound body which consists of I and II in thermal contact so that their temperatures are equal, cf. Sect. 2.1. While the temperatures in both partial bodies are thus equal, the strain fields in I and II are independent. The first law appropriate for a quasistatic process $\vartheta(t), V_\alpha^I(t), V_\alpha^{II}(t)$ in the compound body reads

$$\frac{dU}{dt} + \sum_{\alpha=1}^{n} p_\alpha^I \frac{dV_\alpha^I}{dt} + \sum_{\alpha=1}^{m} p_\alpha^{II} \frac{dV_\alpha^{II}}{dt} = 0, \quad \text{where} \tag{2.13}$$

$$U = U^I\left(\vartheta, V_\alpha^I\right) + U^{II}\left(\vartheta, V_\alpha^{II}\right) \tag{2.14}$$

is the internal energy of the compound body which is assumed to be additive. The inaccessibility axiom applied to the compound body implies the existence of integrating factors Λ such that we have

$$dS_\Lambda = \Lambda \left(dU + \sum_{\alpha=1}^{n} p_\alpha^I dV_\alpha^I + \sum_{\alpha=1}^{m} p_\alpha^{II} dV_\alpha^{II} \right), \text{ or by (2.12), (2.14)} \tag{2.15}$$

$$dS_\Lambda = \frac{\Lambda}{\Lambda^I} dS_\Lambda^I + \frac{\Lambda}{\Lambda^{II}} dS_\Lambda^{II}. \tag{2.16}$$

Therefore S_Λ, hence Λ/Λ^I and Λ/Λ^{II} depend on two variables only. We see that

$$\Lambda/\Lambda^I \quad \text{and} \quad \Lambda/\Lambda^{II} \quad \text{may depend on } S_\Lambda^I \text{ and } S_\Lambda^{II}. \tag{2.17}$$

From this we shall prove that Λ, Λ^I, and Λ^{II} may be written in the forms

$$\Lambda = \frac{\Psi\left(S_\Lambda^I, S_\Lambda^{II}\right)}{T(\vartheta)}, \quad \Lambda^I = \frac{\Psi^I\left(S_\Lambda^I\right)}{T(\vartheta)}, \quad \Lambda^{II} = \frac{\Psi^{II}\left(S_\Lambda^{II}\right)}{T(\vartheta)}, \tag{2.18}$$

where Ψ, Ψ^I and Ψ^{II} are arbitrary functions and $T(\vartheta)$ is a universal function of ϑ, i.e. in particular, it is the same function for the bodies I and II.

For proof of (2.18) we consider this:
By (2.12) $\frac{\Lambda^I}{\Lambda^{II}}$ may be a function of

$$\begin{array}{cc} \vartheta, V_\alpha^I \ (\alpha = 1, 2, ..n) \\ \vartheta, V_\alpha^{II} \ (\alpha = 1, 2, ..m) \end{array}, \text{ or of } \begin{array}{cc} \vartheta, S_\Lambda^I, V_\alpha^I \ (\alpha = 2, ..n) \\ \vartheta, S_\Lambda^{II}, V_\alpha^{II} \ (\alpha = 2, ..m) \end{array}. \tag{2.19}$$

By (2.15) Λ may be a function of

$$\begin{array}{cc} \vartheta, V_\alpha^I \ (\alpha = 1, 2, ..n) \\ \text{and } V_\alpha^{II} \ (\alpha = 1, 2, ..m) \end{array}, \text{ or of } \begin{array}{cc} \vartheta, S_\Lambda^I, S_\Lambda^{II}, V_\alpha^I \ (\alpha = 2, ..n) \\ \text{and} \quad V_\alpha^{II} \ (\alpha = 2, ..m) \end{array}. \tag{2.20}$$

However, from (2.17) and (2.19)$_{1,2}$ we see that

Λ may be a function of $\vartheta, S_\Lambda^I, S_\Lambda^{II}, V_2^I...V_n^I$, or of $\vartheta, S_\Lambda^I, S_\Lambda^{II}, V_2^{II}...V_m^{II}$.

Therefore, since V_α^I $(\alpha = 2, ..n)$ and V_α^{II} $(\alpha = 2, ..m)$ are independent variables, we must conclude that

$$\Lambda \text{ can depend only on } \vartheta, S_\Lambda^I, S_\Lambda^{II}. \tag{2.21}$$

Once again by (2.17), Λ^I and Λ^{II} may then depend on $\vartheta, S_\Lambda^I, S_\Lambda^{II}$ and, since S_Λ^I cannot affect body II, nor S_Λ^{II} body I, we obtain that

$$\begin{matrix} \Lambda^I \\ \Lambda^{II} \end{matrix} \text{ may only depend on } \begin{matrix} \vartheta, S_\Lambda^I \\ \vartheta, S_\Lambda^{II} \end{matrix}.$$

By (2.17) the ϑ-dependence must drop out of the quotients Λ/Λ^I and Λ/Λ^{II}. Therefore it must be given by the same multiplicative function of ϑ in Λ, Λ^I, and Λ^{II}, which we call $\frac{1}{T(\vartheta)}$. This proves (2.18).

With (2.18) we may write (2.12) for either body I or II as

$$\frac{1}{\Psi(S_\Lambda)} dS_\Lambda = \frac{1}{T(\vartheta)} \left(dU + \sum_{\alpha=1}^{n} p_\alpha dV_\alpha \right) \tag{2.22}$$

or with

$$S = \int \frac{dS_\Lambda}{\Psi(S_\Lambda)} + const$$

$$dS = \frac{1}{T(\vartheta)} \left(dU + \sum_{\alpha=1}^{n} p_\alpha dV_\alpha \right). \tag{2.23}$$

Thus, indeed, one of the integrating factors is a universal function of ϑ, viz. $T(\vartheta)$. The corresponding "potential" is called the entropy.

It follows from the additivity of U, cf. (2.14), that the entropy S of two bodies in thermal contact is also additive. Indeed, we have

$$d\left(S^I + S^{II}\right) = \frac{1}{T(\vartheta)} \left(dU + \sum_{\alpha=1}^{n} p_\alpha^I dV_\alpha^I + \sum_{\alpha=1}^{m} p_\alpha^{II} dV_\alpha^{II} \right). \tag{2.24}$$

[Note that the original S_Λ's were not additive, since the integrating factors were different]. Carathéodory says on this occasion: "The additivity of entropy has motivated the physicists to consider the entropy as a quantity that belongs to a body — like its mass — and that depends on the state of the body."

2.6 Absolute (universal) temperature

Equation (2.23) with $U(\vartheta, V_\beta)$ and $p_\alpha = p_\alpha(\vartheta, V_\beta)$ implies integrability conditions, in particular

$$\frac{d\ln T}{d\vartheta} = \frac{\left(\frac{\partial p_\alpha}{\partial \vartheta}\right)_{V_\beta}}{\frac{\partial U}{\partial V_\alpha} + p_\alpha} \qquad (\alpha = 1, 2, ...n). \tag{2.25}$$

It follows that, while p_α and U depend on ϑ and V_β, the combination on the right hand side of (2.25) – for each α – depends only on ϑ. And what is more, *the dependence of those combinations on ϑ is universal.*

The universal character of $T(\vartheta)$ permits us to calculate the form of that function from (2.25) and from the knowledge of the equations of state $p_\alpha(\vartheta, V_\beta)$ and $U(\vartheta, V_\beta)$ of just *one* body. For any ideal gas we have that knowledge, cf. (2.1), where we recall that $f(\vartheta)$ is a known, monotonically increasing function of ϑ. For a gas n is equal to 1 and the only "strain" is the volume V itself. Also there is only one pressure p.

We insert (2.1) into (2.25) and obtain

$$\frac{\mathrm{d}\ln T}{\mathrm{d}\vartheta} = \frac{\mathrm{d}\ln f(\vartheta)}{\mathrm{d}\vartheta}, \quad \text{hence} \quad T(\vartheta) = Cf(\vartheta). \tag{2.26}$$

Thus, to within the multiplicative constant C, the universal function $T(\vartheta)$ is equal to the known function $f(\vartheta)$ that governs the ϑ-dependence of the thermal and caloric equations of state of an ideal gas.

In particular therefore $T(\vartheta)$ is a monotone function of the empirical temperature. Therefore it may itself be used as a measure for temperature. Indeed, William Thomson (Lord Kelvin) proposed to use $T(\vartheta)$ *as the absolute temperature* replacing various empirical temperatures used in his time.[5]

All that is missing is the constant C in (2.26). That constant is determined by fixing the absolute temperature scale and setting the absolute temperature of the triple point of water equal to 273.16 K.

$$T(\vartheta_{Tr^{H_2O}}) = 273.16\,\text{K}. \tag{2.27}$$

Since $f(\vartheta_{Tr^{H_2O}})$ is known – or was known when the empirical temperature ϑ was used – C can be determined from $(2.26)_2$ and what results is

$$\frac{1}{C} = 1.38 \cdot 10^{-23}\,\text{J/K}. \tag{2.28}$$

This value is called the Boltzmann constant, and is denoted by k. It is a universal constant connected with the assignment of the temperature scale.

We may now use k and T to rewrite the thermal and caloric equations of state in their modern form, viz.

$$p = \frac{m}{V}\frac{k}{\mu}T \quad \text{and} \quad U = mz\frac{k}{\mu}(T - T_R) + U(T_R). \tag{2.29}$$

This is the form which we have anticipated in Chap. 1.

---

[5] This reference to Kelvin is somewhat anachronistic, because Kelvin antedates Carathéodory. Kelvin used the universal character of the efficiency of a Carnot engines for an analogous argument to the present one. We have omitted the original Kelvin argument in Chap. 1, because in that chapter we took the absolute temperature for granted.

2.7 Growth of entropy

Inspection of (2.23) shows that S is constant in a reversible adiabatic process, and we may ask what happens to S in an irreversible process. In order to answer that question we consider an irreversible process that begins and ends reversibly so that the integration of the 1st law gives

$$U(S^f, V_\alpha^f) - U(S^i, V_\alpha^i) = -\sum_{\alpha=1}^{n} \int_{t^i}^{t^f} \mathfrak{p}_\alpha \left[\vartheta\,(t-s), V_\beta\,(t-s)\right] \frac{dV_\alpha}{dt} dt. \quad (2.30)$$

Given the value of the work integral this equation may be considered as an equation for S^f. And for different values of the work integral we obtain different values for S^f. Carathéodory assumes that all possible values of S^f lie in a connected interval[6], and now the question arises whether S^i lies in that interval as well. The answer is rather obviously: Yes! Indeed, among all work integrals there are those for reversible processes, which do not affect the entropy.

The next question is whether S^i may lie in an interior point of the connected interval of S^f's; here the answer is No! Indeed, if it did, it would be possible to reach any S^f higher or lower than S^i by a suitable work integral and then continue reversibly to any desired strains V_α^f without changing S. Thus it would be possible to reach any state in the neighbourhood of S^i, V_α^i which, however, is forbidden by the inaccessibility axiom.

We must conclude therefore that S^i lies in one of the two end points of the interval of the S^f's. Actually S^i must lie at the lower end, because in a process from ϑ^i, V_α^i to ϑ^f, V_α^i we must have $\vartheta^f \geqslant \vartheta^i$ − as we saw before in Sect. 2.3 − hence $U^f \geqslant U^i$ by (2.8), and hence $S^f \geqslant S^i$ by (2.23). Therefore *in an adiabatic process the entropy cannot decrease.*

$$\frac{dS}{dt} \geqslant 0. \quad (2.31)$$

Thus Carathéodory comes to the same conclusion as Clausius as far as adiabatic processes are concerned. There is nothing Carathéodory can say about the entropy growth in a non-adiabatic process. His prejudice against heat and heating prevents any such statement.

---

[6] That seems to be a weak assumption. Indeed we should be surprised, if there were a realistic alternative.

3

Entropy in the work of Boltzmann

3.1 Distribution function and Boltzmann equation

We review Boltzmann's work as far as it concerns entropy. Boltzmann was one of the genius physicists who worked on the "mechanical theory of heat" in the 19th century and he found an inequality which he related to the entropy inequality. That inequality should not have occured at all in a mechanical theory and to this day − to our knowledge − no physicists can put the finger on exactly the spot in Boltzmann's chain of arguments where the inequality comes from. And yet, it is generally accepted that − however it was arrived at − Boltzmann's interpretation and extrapolation led to the correct formula $S = k\ln W$ which embodies the so-called statistical definition of entropy. This formula − engraved on Boltzmann's tombstone − is easily the second most important equation in all physics.

Despite the efforts of Clausius and Carathéodory the entropy remained a somewhat implausible concept of little suggestive meaning. Clearly what was needed was a kinetic theory: an interpretation of thermodynamics − and of entropy − in term of atoms and molecules. And that is why many of the foremost scientists in the 1860's and 1870's worked on the *mechanical theory of heat* as the kinetic theory of gases was then called. Clausius, Maxwell and Boltzmann all decided that for a monatomic gas the distribution function $f(\mathbf{x}, \mathbf{c}, t)$ was the most appropriate description of state.

$$f(\mathbf{x}, \mathbf{c}, t)\mathrm{dxdc} \quad - \quad \begin{array}{l} \text{is the number of atoms in the volume element} \\ \text{dx at } \mathbf{x} \text{ with velocities between } \mathbf{c} \text{ and } \mathbf{c} + \mathrm{dc}. \end{array} \quad (3.1)$$

Boltzmann succeeded in writing a differential equation for the distribution function, an equation which is now called Boltzmann equation. This equation forms the basis of the kinetic theory of gases.

Ignoring body forces and inertial forces we may write the Boltzmann equation in the form[1]

---

[1] For the derivation of the Boltzmann equation we refer the reader to any textbook on the kinetic theory of gases. References may be found at the end of this book.

$$\frac{\partial f}{\partial t} + c_i \frac{\partial f}{\partial x_i} = \int \left(f' f^{1\prime} - f f^1 \right) \sigma g \sin \vartheta d\vartheta d\varphi d\mathbf{c}^1. \tag{3.2}$$

This is basically a continuity equation in the space spanned by the coordinates $\mathbf{x}$ and velocities $\mathbf{c}$ of the atoms. The right hand side is due to collisions of atoms with velocities $\mathbf{c}$ and $\mathbf{c}^1$ which, after the collision, have velocities $\mathbf{c}'$ and $\mathbf{c}^{1\prime}$. The angle φ identifies the plane of the binary interaction, while ϑ is related to the angle of deflection of the path of an atom during the collision. σ is the cross-section for a (ϑ, φ)-collision and g is the relative speed of the colliding atoms. The f's in the collision integral are the values of the distribution function for the velocities $\mathbf{c}'$, $\mathbf{c}^{1\prime}$ and $\mathbf{c}$, $\mathbf{c}^1$ respectively as indicated.

It is clear from the definition (3.1) of f that

$$\rho = \int \mu f d\mathbf{c} \qquad \text{– is the mass density of the gas,} \tag{3.3}$$
$$\rho v_i = \int \mu c_i f d\mathbf{c} \qquad \text{– is the momentum density of the gas,}$$
$$\rho \left(u + \tfrac{1}{2} v^2 \right) = \int \tfrac{\mu}{2} \mu c^2 f d\mathbf{c} \quad \text{– is the energy density of the gas.}$$

> The energy of an atom at rest is set equal to zero in the kinetic theory of gases.

The velocity $\mathbf{v}$ of the gas is defined by $(3.3)_2$ and it may be used to introduce the *relative* velocity $\mathbf{C} = \mathbf{c} - \mathbf{v}$ of the atom, the velocity of the atom relative to the moving gas. Thus the internal energy density ρu in $(3.3)_3$ may be separated from the kinetic energy density $\frac{\rho}{2} v^2$ of the gas. We have

$$\rho u = \int \frac{\mu}{2} C^2 f d\mathbf{c}. \tag{3.4}$$

3.2 Equations of transfer

By (3.3) the macroscopic quantities mass, momentum, and energy density have been defined as moments of the distribution function. Therefore the idea suggests itself that, maybe, the Boltzmann equation implies the macroscopic balance equations, or conservation laws, for those quantities. This is indeed the case, as we shall see. The master formula is the equation of transfer for the generic quantity

$$\Psi = \int \int \psi (\mathbf{x}, \mathbf{c}, t) f d\mathbf{c} d\mathbf{x},$$

where $\psi (\mathbf{x}, \mathbf{c}, t)$ is any function with appropriate smoothness and integrability properties. The equation of transfer results from the Boltzmann equation by multiplication with $\psi (\mathbf{x}, \mathbf{c}, t)$ and integration over all velocities. It may be written in the form[2]

---

[2] Once again we refer to books on the kinetic theory of gases for details of the derivation of this equation, in particular in regard to the right hand side.

$$\frac{\partial \int \psi f d\mathbf{c}}{\partial t} + \frac{\partial \left[\int \psi f d\mathbf{c} v_k + \int C_k \psi f d\mathbf{c} \right]}{\partial x_k} - \int \left(\frac{\partial \psi}{\partial t} + c_k \frac{\partial \psi}{\partial x_k} \right) f d\mathbf{c} =$$

$$= \frac{1}{4} \int \left(\psi + \psi^1 - \psi' - \psi_1^{1\prime} \right) \left(f' f^{1\prime} - f f^1 \right) \sigma g \sin \vartheta d\vartheta d\varphi d\mathbf{c}^1 d\mathbf{c}.$$
$$(3.5)$$

This equation has the form of a balance law for a generic quantity Ψ with

density $\qquad \int \psi f d\mathbf{c}$,

convective flux $\qquad (\int \psi f d\mathbf{c}) v_k$,

non-convective flux $\int C_k \psi f d\mathbf{c}$,

intrinsic source $\quad \int \left(\frac{\partial \psi}{\partial t} + c_k \frac{\partial \psi}{\partial x_k} \right) f d\mathbf{c}$, and

collision source $\quad \frac{1}{4} \int \left(\psi + \psi^1 - \psi' - \psi_1^{1\prime} \right) \left(f' f^{1\prime} - f f^1 \right) \sigma g \sin \vartheta d\vartheta d\varphi d\mathbf{c}^1 d\mathbf{c}.$
$$(3.6)$$

ψ^1, ψ' and $\psi^{1\prime}$ stand for $\psi \left(\mathbf{x}, \mathbf{c}^1, t \right)$, $\psi \left(\mathbf{x}, \mathbf{c}', t \right)$ and $\psi \left(\mathbf{x}, \mathbf{c}^{1\prime}, t \right)$, where $\mathbf{c}$, $\mathbf{c}^1$ and $\mathbf{c}'$, $\mathbf{c}^{1\prime}$ are the velocities of the colliding pair of atoms before and after the collision.

We set $\psi = \mu$, $\psi = \mu c_i$ and $\psi = \frac{\mu}{2} c^2$ and conclude that the intrinsic source vanishes for all three cases, because there is no t- nor $\mathbf{x}$-dependence in these ψ's. Also the collision source vanishes, since mass, momentum and energy of the colliding pair of particles are equal before and after the collision. Thus we obtain conservation laws of

mass $\qquad \dfrac{\partial \rho}{\partial t} + \dfrac{\partial \rho v_k}{\partial x_k} = 0,$

momentum $\dfrac{\partial \rho v_i}{\partial t} + \dfrac{\partial \left[\rho v_i v_k + \int \mu C_k C_i f d\mathbf{c} \right]}{\partial x_k} = 0,$ and

energy $\qquad \dfrac{\partial \rho \left(u + \frac{1}{2} v^2 \right)}{\partial t} + \dfrac{\partial \left[\rho \left(u + \frac{1}{2} v^2 \right) v_k + \int \mu C_k C_i f d\mathbf{c} v_i + \int \frac{\mu}{2} C^2 C_k f d\mathbf{c} \right]}{\partial x_k} = 0.$
$$(3.7)$$

In form and physical interpretation these equations correspond to the continuity equation, the equation of motion and the 1st law of continuum mechanics and thermodynamics. Those laws are universally true but here they have the form appropriate to monatomic ideal gases. We recognize that the two remaining integrals in (3.7) — representing the non-convective fluxes of momentum and energy − must be identified as

$$\text{pressure tensor } p_{ki} = \int \mu C_k C_i f d\mathbf{c}, \text{ and}$$

$$\text{heat flux} \quad q_k = \int \frac{\mu}{2} C^2 C_i f d\mathbf{c}. \tag{3.8}$$

While these formulae are useful for the physical understanding of the "mechanisms" of frictional forces and of heat conduction in a gas, they are not of immediate practical use, because the form of the distribution function is unknown.

3.3 Pressure and absolute temperature

However, the pressure p itself − which is equal to $\frac{1}{3}p_{ii}$ − is related, by (3.8) and (3.4), to the internal energy density ρu by the equation

$$p = \frac{2}{3}\rho u, \tag{3.9}$$

irrespective of the form of the distribution function. The relation is obviously satisfied by the thermal and caloric equations of state (1.7) − with $z = \frac{3}{2}$ appropriate for a monatomic gas and if the energy constant is ignored[3]. This observation leads to a kinetic interpretation of the absolute temperature as a measure for the mean kinetic energy of the atoms. Indeed, with $p = \rho\frac{k}{\mu}T$ and (3.4), (3.9) we have

$$\frac{3}{2}kT = \frac{\int \frac{\mu}{2}C^2 f d\mathbf{c}}{\int f d\mathbf{c}}. \tag{3.10}$$

That T is thus practically synonymous with the mean kinetic energy of the atoms or molecules of a body is true for all bodies, not only for monatomic ideal gases. The fact is proved in statistical mechanics − not here (!) − by the so-called equipartition theorem of energy. This interpretation of temperature is intuitively quite easy to grasp, and so it is that

$$\frac{3}{2}kT = \frac{\int \frac{\mu}{2}C^2 f d\mathbf{c}}{\int f d\mathbf{c}} \tag{3.11}$$

is considered as the *definition* of T by many. So suggestive is the new definition that it has largely blanked out the more subtle original definition by which temperature was the quantity which has equal value in two bodies in thermal contact; we recall the discussion of temperature in Sect. 2.1 and 2.6, and we shall come back to the point in Chap. 15.

---

[3] It is clear from $(3.3)_3$ and (3.4) that the kinetic theory ignores the energy of atoms which are at rest.

3.4 Entropy

A candidate for entropy

The ease with which we have derived the conservation laws (3.7) of continuum mechanics and thermodynamics from the generic equation of transfer (3.5) is deceptive. Indeed, for any other choice of $\psi(\mathbf{x}, \mathbf{c}, t)$ – other than μ, μc_i, $\frac{\mu}{2}c^2$ – the collision integral does not vanish, because mass, momentum and energy are the *only* collisional invariants. Therefore it is impossible to obtain more conservation laws.

But there is *one* choice of $\psi(\mathbf{x}, \mathbf{c}, t)$ – discovered by Boltzmann – for which the collision source, although not zero, is non-negative, thus suggesting that it represents the entropy production, or dissipative entropy source. By inspecting $(3.6)_5$ one may actually identify that choice with a little effort – and with the knowledge of hindsight.[4] It reads

$$\psi(\mathbf{x}, \mathbf{c}, t) = a \ln \frac{f}{b}, \tag{3.12}$$

where a is a negative constant to be determined and b is an arbitrary constant needed to make the argument of the logarithm dimensionless.

With (3.12) the collision source has the form

$$\sigma = -\frac{a}{4} \int \ln \frac{f'f^{1\prime}}{ff^1} \left(f'f^{1\prime} - ff^1\right) \sigma g \sin \vartheta d\vartheta d\varphi d\mathbf{c}^1 d\mathbf{c}. \tag{3.13}$$

We have denoted it by σ and obviously σ is non-negative. Thus Boltzmann's kinetic theory identifies the quantity

$$\Psi = \int \int a \ln \frac{f}{b} f d\mathbf{c} d\mathbf{x} \quad \text{with the density} \quad \int a \ln \frac{f}{b} f d\mathbf{c} \tag{3.14}$$

as a quantity with a non-negative source. Therefore there is an immediate suspicion or conclusion: *This quantity may be related to the entropy* or, if a is chosen properly, it may *be* the entropy. Let us investigate that proposition.

First of all, so far, when the entropy has appeared, it has appeared for slow processes, called reversible by Clausius or quasistatic by Carathéodory. Such processes, which are characterized by homogeneous and very slowly changing fields in the body, are also called equilibrium processes and their entropy $S(T, V)$ is therefore an equilibrium entropy; no non-equilibrium entropy has occured up to this point! One of the equilibrium entropies, the one appropriate for ideal gases was given in (1.20). We replace V in (1.20) by $\frac{m}{\rho}$ and absorb m into the constant. Thus we obtain for the equilibrium entropy density of a monatomic gas

---

[4] This was more difficult for Boltzmann. Neither could he employ hindsight, nor did he have the elegant form $(3.6)_5$ for the collision source of a generic quantity.

$$\rho \left[\frac{k}{\mu} \ln \frac{T^{3/2}}{\rho} + const \right]. \tag{3.15}$$

On the other hand, in the kinetic-theory-equation (3.14) nothing refers to equilibrium; f is arbitrary in (3.14) and, if $\int a \ln \frac{f_E}{b} f_E dc$ is to be identified with the equilibrium entropy density (3.15), we first need to determine the equilibrium form f_E of f. We proceed to do so.

Maxwell distribution for equilibrium

Since the entropy remains constant in reversible, or quasistatic, or equilibrium adiabatic processes, we expect the entropy source to vanish. Now, by (3.13), the only distribution function for which the prospective entropy source σ vanishes is such that $\ln f$ is a collisional invariant. That means that $\ln f_E$ must satisfy the relation

$$\ln f_E + \ln f_E^1 = \ln f_E^{\prime} + \ln f_E^{1\prime}. \tag{3.16}$$

However, as mentioned before, there are only five collisional invariants, namely μ, μc_i, $\frac{\mu}{2} c^2$. Therefore $\ln f_E$ must be a linear combination of those

$$\ln f_E = \alpha \mu + \beta_i \mu c_i + \gamma \frac{\mu}{2} c^2.$$

The coefficients α, β_i and γ may be determined from the five moment equations (3.3) which every distribution function must satisfy. Thus we obtain

$$f_E = \frac{\rho/\mu}{\sqrt{2\pi \frac{k}{\mu} T}^3} \; e^{-\frac{\mu(c-v)^2}{2kT}} \;, \quad \text{where} \quad \frac{3}{2} \frac{k}{\mu} T = u. \tag{3.17}$$

This is the well-known Maxwell distribution function. It represents the distribution function in equilibrium.

We insert the Maxwellian f_E into $(3.14)_2$ and obtain

$$\int a \ln \frac{f_E}{b} f_E dc = \rho \left[-\frac{a}{\mu} \ln \frac{T^{3/2}}{\rho} + \left\{ -\frac{a}{\mu} \left(\frac{3}{2} + \ln \left(b\mu \sqrt{2\pi \frac{k}{\mu}}^3 \right) \right) \right\} \right]. \tag{3.18}$$

Thus, if we identify $-a$ with the Boltzmann constant k, $\int a \ln \frac{f_E}{b} f_E dc$ becomes identical with the equilibrium entropy density (3.15), at least in what concerns the ρ- and T-dependence. The constant in curly brackets is indeterminate, because nothing is said about b. And of course in the work of Clausius and Carathéodory the constants are not determined either.

Therefore, we conclude that

$$\rho s = -k \int \ln \frac{f}{b} f dc \quad \text{with the equilibrium value}$$

$$\rho s_E = \rho \left[\frac{5}{2} \frac{k}{\mu} \ln \frac{T}{T_R} - \frac{k}{\mu} \ln \frac{p}{p_R} + \left\{ \frac{k}{\mu} \ln \left(\frac{(kT_R)^{5/2}}{p_R} e^{3/2} \frac{b}{\mu^3} \sqrt{2\pi\mu}^3 \right) \right\} \right] \tag{3.19}$$

may properly be called the entropy density of a gas: It has a positive source for all non-equilibrium distribution functions and its value in equilibrium is identical to the equilibrium entropies of Clausius and Carathéodory.

Note that Boltzmann's entropy $(3.19)_1$ is a quantity vastly more general than the entropies of Clausius and Carathéodory because *Boltzmann's entropy is not restricted to equilibrium.*

Of course, it might seem at this stage that Boltzmann's entropy is restricted to gases. This, however, is not really the case. We shall presently show, how the Boltzmann entropy

$$S = -k \int \int \ln \frac{f}{b} f \mathrm{dcdx} \qquad (3.20)$$

can be, and has been, extrapolated away from gases.

Before that is done, however, we stick to gases and investigate the relation of Boltzmann's entropy to heating.

3.5 Entropy flux, the Clausius-Duhem inequality

We insert $\psi = -k \ln \frac{f}{b}$ into the equation of transfer (3.5) and observe that the intrinsic production vanishes so that the entropy balance assumes the form, cf. (3.19) and (3.13)

$$\frac{\partial \rho s}{\partial t} + \frac{\partial \left[\rho s v_k - k \int C_k \ln \frac{f}{b} f \mathrm{dc} \right]}{\partial x_k} = \sigma \geq 0. \qquad (3.21)$$

This is the local form of a balance whose integral form for a closed system reads[5]

$$\frac{\mathrm{d}S}{\mathrm{d}t} + \int_{\partial V} \Phi_k n_k \mathrm{d}A \geq 0, \quad \text{where} \qquad (3.22)$$

$$\Phi_k = -k \int C_k \ln \frac{f}{b} f \mathrm{dc} \quad \text{is the non-convective entropy flux.} \qquad (3.23)$$

At first sight there is disappointment, if we had hoped to regain Clausius' second law (1.24) in the kinetic theory of gases. Indeed, there is no obvious relation of the flux term in (3.22) to heating, nor does the term contain the temperature. So, does the kinetic theory not confirm the 2nd law? Well, we shall have to go into that question in depth later. However, here we present a preliminary superficial argument which will help to bring the dilemma between the 2nd law and the kinetic theory into sharp focus. Let us consider:

---

[5] There is an appendix at the end of the book for the benefit of those readers who are unfamiliar with the various forms of balance equations.

We write, without loss of generality

$$f = f_E(1 + \varphi), \tag{3.24}$$

where φ represents the deviation of the distribution function from its equilibrium form f_E, cf. (3.17). If this is introduced into (3.23) we find

$$\Phi_k = \frac{q_k}{T} - k \int C_k \ln(1 + \varphi)f d\mathbf{c}, \tag{3.25}$$

where q_k is the heat flux, cf. (3.8). Thus the leading term, if it were the only one, would make (3.22) look like

$$\frac{dS}{dt} + \int_{\partial V} \frac{q_k n_k}{T} dA \geq 0. \tag{3.26}$$

This is the Clausius-Duhem inequality. Since

$$\dot{Q} = - \int_{\partial V} q_k n_k dA \tag{3.27}$$

is the heating, this inequality would agree with the Clausius inequality (1.24), if the temperature were homogeneous on ∂V. We recall that that was indeed the case considered by Clausius. The Duhem correction in (3.26) is a natural extrapolation of the second law (1.24) to the case where the heated body has a non-homogeneous boundary temperature.

However, the second term in (3.25) does not fit into the Clausius scheme in any obvious manner. We shall later devote a whole chapter to it – albeit a short one, cf. Chap. 15 on the zeroth law of thermodynamics.

3.6 Reformulation of Boltzmann's entropy

We have started this treatise on the kinetic theory with the hope of finding a suggestive interpretation of entropy. And now, after we have found the kinetic expression (3.20) for entropy, we must realize that it does not help us, at least not in that form. Or how would we interpret the mean value of $-k \ln \frac{f}{b}$ suggestively? And yet, with the following sequence of arguments it is possible to hammer (3.20) into a form that *is* suggestive – or nearly so. Let us consider:

We take it for granted that the $N_{d\mathbf{x}d\mathbf{c}}$ atoms in the element $d\mathbf{c}d\mathbf{x}$ occupy discrete points $\mathbf{x}\mathbf{c}$ of which there are $P_{d\mathbf{x}}Q_{d\mathbf{c}}$ in $d\mathbf{c}d\mathbf{x}$, cf. Fig. 3.1. $P_{d\mathbf{x}}$ is the number of positions and $Q_{d\mathbf{c}}$ the number of velocities in the element. Let there be $N_{\mathbf{x}\mathbf{c}}$ atoms on the point $\mathbf{x}\mathbf{c}$. It is then clear from the definition of f that we must have

$$f\,\mathrm{dxdc} = N_{\mathrm{dxdc}} = \sum^{P_{\mathrm{dx}}Q_{\mathrm{dc}}} N_{\mathbf{xc}}, \qquad (3.28)$$

where the summation extends over all points in dxdc.

We assume that there is *local equilibrium*, meaning that within dcdx the distribution is homogeneous so that all of the $P_{\mathrm{dx}}Q_{\mathrm{dc}}$ points in dcdx are occupied by the same number of atoms. This implies

$$N_{\mathbf{xc}} = \frac{N_{\mathrm{dxdc}}}{P_{\mathrm{dx}}Q_{\mathrm{dc}}} \quad \forall\ \mathbf{xc} \in \mathrm{dcdx}. \qquad (3.29)$$

Note that $N_{\mathbf{xc}}$, the number of atoms per point, may well be smaller than 1. We assume that the number of points $P_{\mathrm{dx}}Q_{\mathrm{dc}}$ in the element dcdx is proportional to its size dcdx with XY as factor of proportionality.[6] Thus we have

$$P_{\mathrm{dx}}Q_{\mathrm{dc}} = XY\mathrm{dcdx} \quad \text{and hence, by (3.28), (3.29): } f\,(\mathbf{x},\mathbf{c}) = XYN_{\mathbf{xc}}. \qquad (3.30)$$

We may then write the contribution $S_{\mathbf{xc}}\mathrm{dcdx}$ of the element dcdx to the entropy (3.20) as

$$S_{\mathbf{xc}}\mathrm{dcdx} = -k \sum^{P_{\mathrm{dx}}Q_{\mathrm{dc}}} \ln\left(\frac{XY}{b}N_{\mathbf{xc}}\right) N_{\mathbf{xc}}, \qquad (3.31)$$

which, by (3.29), is really a sum over $P_{\mathrm{dx}}Q_{\mathrm{dc}}$ equal terms.

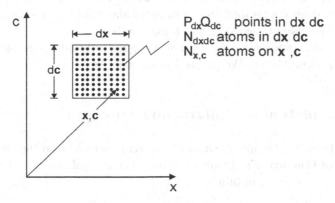

Fig. 3.1 An $(\mathbf{x},\mathbf{c})$-space element dcdx at $\mathbf{x},\mathbf{c}$ with discrete points that may be occupied by atoms.

---

[6] We consider the number of positions in dx as $X\mathrm{dx}$ and the number of velocities in dc as $Y\mathrm{dc}$. Until further notice we may consider the implied discreteness of positions and velocities as a heuristic device. Later, in Chap.14 we shall see that XY is related to the Planck constant.

We recall that b was introduced as an arbitrary factor of the dimension $\frac{s^3}{m^6}$. We may choose it at will and now is the time to get rid of it. We set

$$b = eXY \tag{3.32}$$

so that (3.31) assumes the form

$$S_{\mathbf{xc}}\mathrm{d}\mathbf{c}\mathrm{d}\mathbf{x} = -k \sum^{P_{\mathrm{d}\mathbf{x}}Q_{\mathrm{d}\mathbf{c}}} (N_{\mathbf{xc}} \ln N_{\mathbf{xc}} - N_{\mathbf{xc}}). \tag{3.33}$$

Therefore the total entropy turns out to be given by

$$S = -k \sum_{\mathbf{xc}}^{PQ} (N_{\mathbf{xc}} \ln N_{\mathbf{xc}} - N_{\mathbf{xc}}), \tag{3.34}$$

where P is the total number of positions $\mathbf{x}$ in V, and Q the total number of velocities $\mathbf{c}$ in $-\infty < \mathbf{c} < \infty$. That allows us to write S in the compact form

$$S = k \ln \frac{1}{\prod_{\mathbf{xc}}^{PQ} N_{\mathbf{xc}}!}, \tag{3.35}$$

provided that the numbers $N_{\mathbf{xc}}$ are big enough to use the Stirling approximation $\ln a! \approx a \ln a - a$ for large values of a. [Recall the caution below (3.29) that $N_{\mathbf{xc}}$ may be smaller than 1. Clearly that caution is ignored in the transition between (3.34) to (3.35). We shall come back to this point in Chap. 16. For Boltzmann and his contemporaries − in particular Gibbs − it seems to have been natural to assume that $N_{\mathbf{xc}}$ was a big number.]

This new expression for S is still not a suggestively interpretable expression, but it is close to one. We proceed to consider this.

3.7 Extrapolation of Boltzmann's entropy

It is hugely tempting to smuggle a constant term $+k \ln N!$ into the alternative form (3.35) of Boltzmann's entropy, where N is the total number of atoms of the gas. If we do this we obtain[7]

$$S = -k \ln W \quad \text{with} \quad W = \frac{N!}{\prod_{\mathbf{xc}}^{PQ} N_{\mathbf{xc}}!}. \tag{3.36}$$

---

[7] We are told that Boltzmann never wrote $S = k \ln W$, although that formula is engraved on his tombstone, cf. Chap. 21.

With that additional term the argument W of the logarithm may be interpreted — by the rules of combinatorics — as the number of possible realizations of the distribution $\{N_{\mathbf{xc}}\}$, and thus lead to an intuitive understanding of entropy and its growth property.

Indeed, one may plausibly assume that each realization occurs equally frequently in the course of the thermal fluctuations of the gas. If that is true, then the gas, starting with any distribution, will naturally — or most probably — move to distributions with more realizations and eventually it will be seen in *the* distribution with most realizations.

Such an interpretation of S as $k \ln W$ explains the growth of entropy and its trend to a maximum so naturally, that it *simply has to be true*. Therefore any possible uneasiness about smuggling $k \ln N!$ into the correct expression (3.35) must have seemed to 19th century physicists to be a small price to pay.

And yet there was a catch or snag connected with the $N!$ in (3.36). We shall explain that in Chap. 16 when we consider the Gibbs paradox. It will turn out there that, in fact, the correct formula (3.34) itself may be interpreted as $S = k \ln W$ — with W as a number of realizations of a distribution and irrespective of the size of $N_{\mathbf{xc}}$ — but it took quantum mechanics to see that.

Not only was $S = k \ln W$ interpretable, it could also be extrapolated away from gases to any body, or any ensemble consisting of many parts and thus define the entropy of the body or of the ensemble. Take the wealth in a population: Let there be N dollars and let them be distributed over the P persons of a population such that $\{N_1, N_2, ...N_P\}$ is the number of dollars owned by each one. Thus

$$S = k \ln \frac{N!}{\displaystyle\prod_{i=1}^{P} N_i!} \tag{3.37}$$

may be defined as the entropy of the wealth of the population — whatever that is good for.

We shall come to learn of a less facetious example when we treat rubber molecules in Chap. 5. Also it will then become clearer what the significance of a large value of W is and why entropy has the tendency to grow.

3.8 Entropy constant and quantization of $(\mathbf{x}, \mathbf{c})$-space

We use $b = eXY$, cf. (3.32), to rewrite the equilibrium entropy density (3.19) in a monatomic gas in its final form

$$\rho s_E = \rho \left[\frac{5}{2} \frac{k}{\mu} \ln \frac{T}{T_R} - \frac{k}{\mu} \ln \frac{p}{p_R} + \left\{ \frac{k}{\mu} \ln \frac{(kT_R)^{5/2}}{p_R} e^{5/2} \frac{XY}{\mu^3} \sqrt{2\pi\mu}^3 \right\} \right] . \tag{3.38}$$

The entropy constant — the quantity in curly brackets — is still indeterminate, because we do not know XY. What is XY?

By $(3.30)_1$ the smallest element of the $(\mathbf{x}, \mathbf{c})$-space, i.e. the element that can accommodate one occupiable point is $\frac{1}{XY}$. Thus the $(\mathbf{x}, \mathbf{c})$-space is *quantized* into cells of size $\frac{1}{XY}$. And indeed it will turn out − in Chap. 14 − that XY is related to the Planck constant.

4

Enthalpy and equations of state

It is all very well to advertise a concept like entropy as important for natural philosophy, but it is much more efficient to introduce it as a means for making money, or saving money. And, indeed, that is what entropy does! It saves time and money, because it helps to determine the caloric equation of state of gases and fluids. We need to know the form of that equation when we build refrigerators, or when we refine natural gas or oil, or when we produce fertilizers and explosives.

It is often useful and very common to represent the caloric equation of state as an equation for enthalpy rather than internal energy. Enthalpy is a form of energy — appropriate for reversible processes — which combines the internal energy of the body with the potential energy of the loading device, e.g. a heavy piston, that creates and maintains the pressure of the body.

4.1 Enthalpy - a common form of energy

One of the first and foremost consequences of the 1st and 2nd laws for reversible processes is the *Gibbs equation* (1.19) or, for a unit mass[1]

$$
\left.
\begin{aligned}
\dot{q} &= T\frac{ds}{dt} \\
\dot{q} &= \frac{du}{dt} + p\frac{dv}{dt}
\end{aligned}
\right\}
\implies \frac{ds}{dt} = \frac{1}{T}\left(\frac{du}{dt} + p\frac{dv}{dt}\right).
\tag{4.1}
$$

An alternative form of the Gibbs equation makes use of the specific enthalpy $h = u + pv$ instead of the internal energy u

$$
\left.
\begin{aligned}
\dot{q} &= T\frac{ds}{dt} \\
\dot{q} &= \frac{dh}{dt} - v\frac{dp}{dt}
\end{aligned}
\right\}
\implies \frac{ds}{dt} = \frac{1}{T}\left(\frac{dh}{dt} - v\frac{dp}{dt}\right).
\tag{4.2}
$$

---

[1] We follow the usual practice and denote global quantities by capital letters and mass-specific quantities by the corresponding minuscules.

Engineers prefer $h(T,p)$ over $u(T,v)$ as the relevant caloric equation of state, because in the prototypical heat engine of Fig. 1.1$_{left}$ the heating and the working are given by changes of enthalpy. Therefore the practical thermodynamicist's most important aids are the (h,s)-diagram for heat engines and the $(\log p, h)$-diagram for refrigerators; both have enthalpy values on one coordinate axis.

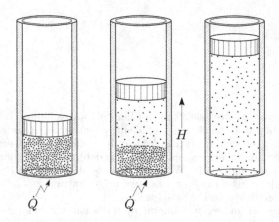

Fig.4.1 Isobaric and isothermal evaporation by heating.

Also experimentalists – particularly chemists – prefer enthalpy over internal energy, because it is easier to control pressure than volume and because volume changes are easier to measure than pressure changes. The prototypical case of a pressure-controlled reversible experiment is shown in Fig. 4.1, where a fixed pressure is maintained by the weight $m_p g$ of a piston with cross-sectional area A. Inspection of (4.2) shows that heating equals the rate of change of enthalpy in such a case – not the rate of change of internal energy – because for $p =$const we may write

$$\dot{q} = \frac{dh}{dt} = \frac{du}{dt} + p\frac{dv}{dt};$$

this implies that the heating transforms itself into internal energy, plus work needed to lift the piston, which increases its potential energy $m_p g H$. Indeed, the term pV in the enthalpy $H = U + pV$ may be rewritten as a potential energy, because we have $p = \frac{m_p g}{A}$ and $V = AH$. Thus we understand that in a reversible isobaric process *enthalpy represents energy of the body plus potential energy of the dead load that produces the pressure.*

In particular, if a phase transition t – from liquid to vapour (say) – occurs isobarically at the pressure p and temperature $T_t = T(p)$, the equation

(4.2) implies that the *latent heat* $r(T_t)$ of the transition $-$ e.g. the heat of evaporation $-$ may be expressed as the change $\Delta h(T_t)$ of enthalpy; of course, it is also related to the change $\Delta s(T_t)$ of entropy. We have

$$r(T_t) = \Delta h(T_t) = T_t \Delta s(T_t). \tag{4.3}$$

4.2 Equations of state

The importance of the Gibbs equation can hardly be overestimated; it saves time and money and it is literally worth billions to the chemical industry, because it reduces drastically the number of measurements, which must be made in order to determine the caloric equation of state $u = u(T, v)$, or $h = h(T, p)$. Let us consider:

Both the thermal equation of state $p = p(T, v)$ and the caloric equation of state $h = h(T, p)$ are needed explicitly for the calculation of nearly all thermodynamic processes and they must be measured. Now, it is easy $-$ at least in principle $-$ to determine the thermal equation, because p, v, and T are all measurable quantities and they need only be put down in tables, or diagrams, or $-$ in modern times $-$ on CD's. But that is not so with the caloric equation, because neither h nor u are measurable quantities. Therefore, in order to determine $h(T, p)$ we must detour via the two specific heat capacities, one for fixed pressure and the other one for fixed volume[2]

$$c_p = \left(\frac{\partial h}{\partial T}\right)_p \quad \text{and} \quad c_v = c_p + \left(\frac{\partial p}{\partial T}\right)_v \left[\left(\frac{\partial h}{\partial p}\right)_T - v\right]. \tag{4.4}$$

These *can* be measured and, once they are known, we may use (4.4) to calculate the derivatives $\left(\frac{\partial h}{\partial T}\right)_p$ and $\left(\frac{\partial h}{\partial p}\right)_T$ of h and hence $-$ by integration $-$ $h(T, p)$ itself to within an additive constant. However, although possible, measurements of $c_p(T, p)$ and $c_v(T, p)$, like all caloric measurements, are notoriously difficult, time-consuming, expensive $-$ and unreliable to boot. And that is where the Gibbs equation helps. We consider this:

We use $h = h(T, p)$ to write the Gibbs equation (4.2) in the form

$$ds = \frac{1}{T}\left(\frac{\partial h}{\partial T}\right)_p dT + \frac{1}{T}\left(\left(\frac{\partial h}{\partial p}\right)_T - v\right) dp. \tag{4.5}$$

This relation implies an integrability condition, viz.

---

[2] A common and useful convention of thermodynamics concerns partial derivatives of functions of two - or more - variable. They carry an index identifying the variable, or variables, that are kept constant. This practice helps in keeping track of frequent changes of variables. Thus for instance, sometimes we need $s(T,v)$ and sometimes $s(T,p)$. Obviously the partial derivatives with respect to T are different in the two cases and they should be denoted differently.

$$\left(\frac{\partial h}{\partial p}\right)_T = v - T \left(\frac{\partial v}{\partial T}\right)_p, \tag{4.6}$$

which relates $\left(\frac{\partial h}{\partial p}\right)_T$ to the *thermal* equation of state. Therefore measurements of c_v, and the equation $(4.4)_2$, are *not* needed for the calculation of $\left(\frac{\partial h}{\partial p}\right)_T$. Moreover from $(4.4)_1$ and (4.6) we obtain

$$\left(\frac{\partial c_p}{\partial p}\right)_T = T \left(\frac{\partial^2 v}{\partial T^2}\right)_p, \tag{4.7}$$

so that the p-dependence of c_p is also related to the thermal equation of state. All that remains to be measured calorically is c_p for *one* p as a function of T. Therefore the use of the Gibbs equation has drastically reduced the necessary number of costly and laborious caloric measurements for the calculation of $h(T, p)$. And that, for practical purposes, is perhaps the most important consequence of the 2nd law.

Integration of $(4.4)_1$ and (4.6) furnishes $h(T, p)$ and $s(T, v)$, the latter from the Gibbs equation (4.5), and we obtain

$$h(T, p) = \int_{p_R}^{p} \left[v(T, \alpha) - T \left(\frac{\partial v}{\partial T}\right)_p (T, \alpha) \right] d\alpha +$$

$$+ \int_{T_R}^{T} c_p(\beta, p_R) d\beta + \sum_i r(T_{t_i}) + h(T_R, p_R)$$

$$\tag{4.8}$$

$$s(T, p) = \int_{p_R}^{p} - \left(\frac{\partial v}{\partial T}\right)_p (T, \alpha) d\alpha +$$

$$+ \int_{T_R}^{T} \frac{c_p(\beta, p_R)}{\beta} d\beta + \sum_i \frac{r(T_{t_i})}{T_{t_i}} + s(T_R, p_R).$$

The sums represent the changes of h and s due to latent heats if, along the triangular path from $(T_R, p_R) \to (T, p_R) \to (T, p)$, the body undergoes phase transitions t_i at the temperatures T_{t_i}, cf. (4.3).

Thus $h(T, p)$ and $s(T, p)$ may be calculated from (p, v, T)-measurements, measurements of $c_p(T, p_R)$, and measurements of latent heats $r(T_t)$. Both functions contain additive constants – $h(T_R, p_R)$ and $s(T_R, p_R)$ respectively – whose values are unimportant until later when we begin to consider chemical reactions. T_R and p_R are the usual reference values for T and p, viz. $T_R = 25°$C or 298.15 K and $p_R = 1$ atm.

4.3 Two special cases

Simple and analytically explicit special cases of the equations of state (1.6) include those for ideal gases and for incompressible bodies.

For an ideal gas we know the thermal and caloric equations of state, cf. (1.7)

$$p = \frac{1}{v}\frac{k}{\mu}T,$$

$$h = (z+1)\frac{k}{\mu}(T - T_R) + h(T_R), \qquad \text{hence by (4.8)}_2 \qquad (4.9)$$

$$s = (z+1)\frac{k}{\mu}\ln\frac{T}{T_R} - \frac{k}{\mu}\ln\frac{p}{p_R} + s(T_R, p_R).$$

Inspection shows that h is independent of v, or p, and that s changes logarithmically with both T and p. Of course, (4.9) is only valid, if no phase transition occurs between (T_R, p_R) and (T, p); otherwise a term, or terms, with the latent heat have to be taken into account.

Another simple — idealized — case of considerable heuristic importance is the case of an incompressible solid or liquid with a constant specific heat c.

$$v = const.$$

$$h(T, p) = c(T - T_R) + (p - p_R)v + h(T_R, p_R), \quad \text{hence by (4.8)}_2 \qquad (4.10)$$

$$s(T) = c\ln\frac{T}{T_R} + s(T_R).$$

Thus the enthalpy is a linear function of T and p, while the entropy grows logarithmically with T; it is independent of p. Once again (4.10) holds only, if there is no phase transition between (T_R, p_R) and (T, p).

4.4 Van der Waals equation for a "real" gas

Van der Waals has used statistical arguments to derive a thermal equation of state for a "real" gas — as opposed to an ideal gas. He reasoned that the atoms of a gas interact with a force — now called *van der Waals force* — that is mildly attractive at large distances and strongly repulsive when the particles are close. Thus the potential $\phi(r)$ of the interaction force between two particles in the distance r has a shape shown qualitatively by the graph of Fig. 4.2. On this basis van der Waals was able to derive a modified form of the ideal gas equation, namely

$$p(v, T) = \frac{1}{v - b}\frac{k}{\mu}T - \frac{a}{v^2}. \qquad (4.11)$$

The modification lies in the coefficients a and b, where a represents the size of the range of interaction around an atom, and b represents the actual volume filled by the atom, cf. Fig. 4.2.

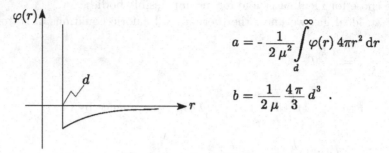

$$a = -\frac{1}{2\,\mu^2}\int_d^\infty \varphi(r)\,4\pi r^2\,dr$$

$$b = \frac{1}{2\,\mu}\frac{4\pi}{3}\,d^3\ .$$

Fig. 4.2 Interatomic interaction potential as a function of the distance of two atoms. Also: van der Waals coefficients.

The statistical arguments of van der Waals were not good enough for the calculation of the caloric equation of state. But we can go a long way to determine $u(T,v)$[3] and $s(T,v)$ by exploiting the Gibbs equation (4.1). We have

$$\left(\frac{\partial u}{\partial T}\right)_v = c_v \quad \text{and} \quad \left(\frac{\partial u}{\partial v}\right)_T = -p + T\left(\frac{\partial p}{\partial T}\right)_v . \tag{4.12}$$

These equations are analogues of $(4.4)_1$ and (4.6) which have made use of the Gibbs equation in the form (4.1). Insertion of the thermal equation of state (4.11) into $(4.12)_2$ and integration of both equations (4.12) gives.

$$u(T,v) = -a\left(\frac{1}{v} - \frac{1}{v(T_R, p_R)}\right) + \int_{T_R}^T c_v(\beta)\mathrm{d}\beta + u\left(T_R, v(T_R, p_R)\right) . \tag{4.13}$$

Note that c_v is independent of v, because p is a linear function of T.

By (4.11) and (4.13) we obtain from the integration of the Gibbs equation (4.1)

$$s(T,v) = \frac{k}{\mu}\ln\frac{v-b}{v(T_R, p_R) - b} + \int_{T_R}^T \frac{c_v(\beta)}{\beta}\mathrm{d}\beta + s\left(T_R, v(T_R, p_R)\right) . \tag{4.14}$$

$c_v(T)$ must be determined from measurements. But, apart from that we now have $u(T,v)$ and $s(T,v)$ to within an additive constant each.

---

[3] For consistency with (4.9) and (4.10) we should have preferred to calculate $h(T,p)$ rather than $u(T,v)$. But that is impractical for the van der Waals gas, because the calculation of $v(T,p)$ requires the solution of a cubic equation which, although possible, is best avoided.

In some strange and surprising manner the van der Waals equations (4.11), (4.13), (4.14) are capable of describing not only real gases and vapours, but also liquids and the phase transition between them. To be sure, the theory is not good enough for any real substance, but its heuristic value for qualitative investigations is extraordinary. We shall demonstrate the usefulness in Sect. 10.5.

5

Gases and Rubber

We demonstrate that the entropy is "driven" toward a maximum by the random character of the thermal motion. That drive is capable of exerting forces. Thus the expansive pressure of a gas and the elastic force of a stretched rubber strap are both *entropic* forces. The study of these cases — gases and rubber — can provide a thorough appreciation of the "mechanism" by which entropy grows. At the same time it becomes clear that the growth is merely probable, albeit *very probable*, but not strictly deterministic.

Entropy is a measure of disorder; this aspect of entropy is best understood by considering a polymeric rubber molecule. The knowledge of entropy of a rubber molecule implies knowledge of the entropy of a rubber strap which, in turn, allows us to calculate the thermal equation of state of rubber. Altogether this chain of arguments is known as the *kinetic theory of rubber*. Different as gases and rubber may be in appearance, thermodynamically those materials are essentially identical. A joker with an original turn of mind has once commented on this similarity by saying that "rubbers are the ideal gases among the solids". Both exhibit entropic elasticity.

5.1 Energetic and entropic parts of elasticity

We recall the 1ˢᵗ and 2ⁿᵈ laws of thermodynamics (1.1) and (1.24) with the working or the power of the stress t_{ij} given by

$$\dot{A} = \int_{\partial V} t_{ij} v_i n_j \mathrm{d}A, \qquad (5.1)$$

and we apply those to the bodies shown in Fig. 5.1: A fluid of volume V under a pressure p and a solid rod of length L under the tensile force P. Both have the homogeneous temperature T and the working is applied slowly so that the processes in the bodies are reversible; the kinetic energy is negligible.

We treat fluids and solids simultaneously — in juxtaposition. Formulae, comments and figures referring to fluids are found on the left hand side of the central dividing line, while the right hand side refers to solids.

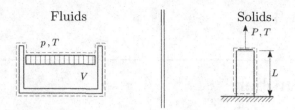

Fig. 5.1 Fluids under the pressure p and
solid rod under tensile load P.

The working effects a change of volume V of the fluid and of length L of the rod. It reads

$$\dot{A} = -p\frac{\mathrm{d}V}{\mathrm{d}t} \quad \| \quad \dot{A} = P\frac{\mathrm{d}L}{\mathrm{d}t}. \tag{5.2}$$

Thus elimination of $\dot{Q}$ between the 1st and 2nd laws gives the Gibbs equation appropriate to the two cases

$$\mathrm{d}S = \tfrac{1}{T}\left(\mathrm{d}U + p\mathrm{d}V\right) \quad \| \quad \mathrm{d}S = \tfrac{1}{T}\left(\mathrm{d}U - P\mathrm{d}L\right). \tag{5.3}$$

Internal energy and pressure p, or load P are given by the caloric and thermal equations of state, viz.

$$U = U(T,V), \ p = p(T,V) \quad \| \quad U = U(T,L), \ P = P(T,L). \tag{5.4}$$

It is thus possible to rewrite the Gibbs equation as an expression involving the *free energy* $F = U - TS$

$$\mathrm{d}(U - TS) = -S\mathrm{d}T - p\mathrm{d}V \ \| \ \mathrm{d}(U - TS) = -S\mathrm{d}T + P\mathrm{d}L \tag{5.5}$$

and it follows that p and P are derivatives of the free energy $U - TS$ with respect to V and L, respectively

$$p = -\left(\tfrac{\partial U}{\partial V}\right)_T + T\left(\tfrac{\partial S}{\partial V}\right)_T \ \| \ P = \left(\tfrac{\partial U}{\partial V}\right)_T - T\left(\tfrac{\partial S}{\partial L}\right)_T \tag{5.6}$$

so that it is possible to say that the pressure and load have two parts: An *energetic* one and an *entropic* one given by the derivatives of U and S with respect to V, or L.

It is desirable − and possible − to read-off those two parts of the load separately from a plot constructed by a simple experiment: The measurement of a (p, T)- or (P, T)-diagram, cf. Fig. 5.2. The experiment consists of heating the body and monitoring the values of p, or P which keep the volume V, or length L constant. Figure 5.2 shows qualitatively the graphs thus produced.

In a given point the slope of the graph identifies the entropic part of the load, because the integrability conditions implied by (5.5) read

$$\left(\tfrac{\partial S}{\partial V}\right)_T = \left(\tfrac{\partial p}{\partial T}\right)_V \, \Big\| \, \left(\tfrac{\partial S}{\partial L}\right)_T = -\left(\tfrac{\partial P}{\partial T}\right)_L. \tag{5.7}$$

Also, by (5.6), (5.7) we have

$$\left(\tfrac{\partial U}{\partial V}\right)_T = -p + T\left(\tfrac{\partial p}{\partial T}\right)_V \, \Big\| \, \left(\tfrac{\partial U}{\partial L}\right)_T = P - T\left(\tfrac{\partial P}{\partial T}\right)_L, \tag{5.8}$$

so that the energetic part of the load is represented by the ordinate intercept of the tangent of the graphs.

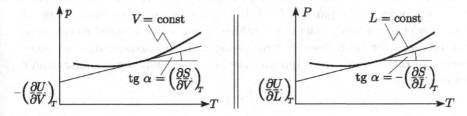

Fig. 5.2 Entropic and energetic parts of pressure or load.

5.2 The elasticity of ideal gases and amorphous rubber is entropy-induced

If the experiments for recording the (p, T)- or (P, T)-diagram are performed for

ideal gases || amorphous rubber

it turns out that the graphs are straight lines which − when extrapolated back to absolute zero − pass through the origin, cf. Fig. 5.3. It is therefore clear that for those materials the internal energy is independent of V, and L and (5.6) reduces to

$$p = T\left(\tfrac{\partial S}{\partial V}\right)_T \, \| \, P = -T\left(\tfrac{\partial S}{\partial L}\right)_T . \tag{5.9}$$

We express this by saying that the pressure p of ideal gases or the force of a rubber strap are *entropy induced*. Meaning the same thing, one also says that the elasticity of gases or rubber is entropic. This property is peculiar to gases and rubber and it reveals a deep-seated similarity − thermodynamically − between these materials, despite their very different appearances.

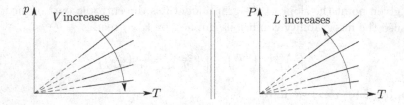

Fig. 5.3 Isochores of an ideal gas and lines of constant length of amorphous rubber pass through the origin.

It is clear from (5.9) that we may calculate the thermal equations of state $p(T,V)$ for ideal gases and $P(T,L)$ for amorphous rubber from the knowledge of the entropies $S(T,V)$ and $S(T,L)$. In the case of gases we already have all the knowledge we need, but not for rubber. Even so, we proceed to calculate the entropies for both cases by continuing with the juxtaposition of arguments. In that manner we emphasize the universal character of Boltzmann's expression

$$S = k \ln W. \tag{5.10}$$

5.3 $S = k \ln W$ for gases and a rubber molecule

We need models for the calculation of W which represents the number of states that can realize a distribution. In the case of a monatomic gas the model is clear: We have N atoms at positions $\mathbf{x}$ with velocities $\mathbf{c}$ in a box of volume V and their total energy is U, cf. Fig. 5.4. In rubber the molecules are in general long entangled chains of N independently oriented links, each of length b, and the chain has an end-to-end distance r. The individual links are isoprene molecules[1] and the rubber chain is a polyisoprene molecule containing thousands of links. Therefore we can ascribe an entropy to an individual molecule.

As before, in Sect. 3.6, we envisage discrete positions $\mathbf{x}$ and velocities $\mathbf{c}$ in V which the atoms of the gas can occupy and similarly we take the orientations ϑ, φ of links of the rubber molecule to be discrete. As before in Sect.3.6 we denote the number of positions by P and the number of velocities by Q. Similarly the numbers of polar and azimuthal angles are denoted by P and Q respectively.

---

[1] We oversimplify. In reality two consecutive isoprene links are not independent in orientation. It takes a group of several such molecules to obtain independence of the orientation between the beginning of the group and the end.

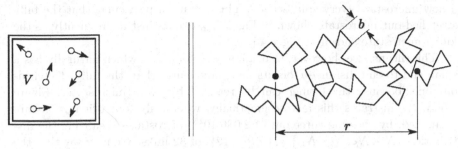

Fig. 5.4 Models of a gas and of a rubber molecule.

A typical realization, often called microstate, may be given by

| in a gas || for a rubber molecule |

| atom i lies at a particular position $\mathbf{x}$ and has a particular velocity $\mathbf{c}$ $(i = 1, 2...N)$ || link i has a particular polar angle ϑ and a particular azimuthal angle φ $(i = 1, 2...N)$. |

The distributions are given by $\{N_{\mathbf{xc}}\}$ and $\{N_{\vartheta\varphi}\}$ respectively, where $N_{\vartheta\varphi}$ denotes the number of links with orientation ϑ, φ. It is then clear that W, the number of microstates that can realize a distribution is given by

$$W = \frac{N!}{\prod\limits_{\mathbf{xc}}^{PQ} N_{\mathbf{xc}}!} \qquad\qquad W = \frac{N!}{\prod\limits_{\vartheta\varphi}^{PQ} N_{\vartheta\varphi}!}$$

$$\qquad\qquad\qquad\qquad\qquad\qquad (5.11)$$

$$W = -\sum_{\mathbf{xc}}^{PQ} N_{\mathbf{xc}} \ln \frac{N_{\mathbf{xc}}}{N} \qquad\qquad W = -\sum_{\vartheta\varphi}^{PQ} N_{\vartheta\varphi} \ln \frac{N_{\vartheta\varphi}}{N},$$

where the Stirling formula has been employed to obtain the second line. Hence follow the entropies $S = k \ln W$.

5.4 Growth of entropy, entropic forces

The rubber molecule, viewed as a chain of independent links, is the most instructive model for understanding the growth of entropy. The basic idea is the concept of equal *a priori* probability of each microstate, by which each and every microstate occurs just as frequently as any other one during the course of the thermal motion. This is the only reasonable unbiased assumption given the random character of the thermal motion that kicks the chain into a new microstate every split second. This means in particular that the fully stretched-out microstate shown in Fig. 5.5$_{\text{top}}$ occurs just as frequently as the kinky microstate shown in Fig. 5.4$_{\text{right}}$.

This means also, however, that a kinky *distribution* which naturally has a small end-to-end distance r occurs more frequently than the fully stretched-out one, because the former can be realized by more microstates. Figure 5.5$_{\text{bottom}}$ illustrates this fact schematically, when only four orientations are permitted, by showing three out of $2.086 \cdot 10^{15}$ microstates[2] realizing the distribution $\{N_N, N_S, N_W, N_E\} = \{9, 9, 2, 12\}$ of 32 links. We may say that the end-to-end distance r shown in Fig. 5.5$_{\text{bottom}}$ occurs $2.086 \cdot 10^{15}$ times more often than the end-to-end distance $r = Nb$ of the stretched-out chain at the top.

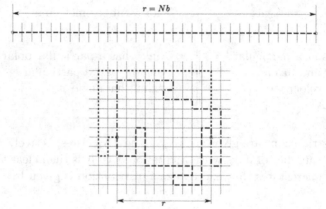

Fig. 5.5 Top: Stretched-out chain of 32 links
Bottom: 3 out of $2.086 \cdot 10^{15}$ possibilities
to realize the distribution $\{9, 9, 2, 12\}$.

Most frequent is *the* distribution with the most realizations and, rather obviously, that distribution is the isotropic one, where all $N_{\vartheta\varphi}$ are equal, so that the end-to-end distance is zero. For many links – large N – the number of microstates realizing that isotropic distribution is bigger, or much bigger than the number of microstates of all other distributions.

Therefore, if the molecule starts out straight – with $W = 1$, i.e. $S = 0$ – the thermal motion will very quickly mess up this straight distribution and kick the molecule into a kinky distribution and – eventually – with overwhelming probability, into the isotropic distribution of maximum W, i.e. maximum S, which we call the equilibrium distribution; this distribution does not change anymore except for the inevitable fluctuations associated with the thermal motion. That is the nature of the growth of entropy as the molecule tends to equilibrium.

---

[2] This large number is the calculated value of $\frac{32!}{9!9!2!12!}$.

Can we prevent this growth of entropy? Yes, we can! If we wish to keep the molecule straight or with a large end-to-end distance r, we only need to give a tug at the ends each time the thermal motion kicks it. And, if the thermal motion kicks the molecule 10^{12} times per second, we may apply a constant force at the ends. That is the nature of entropic forces and of entropic elasticity. And that is the nature of the force needed to keep a rubber strap extended.

Of course, the thermal motion does not stop when the equilibrium is reached. It continues to kick the molecule around. Therefore the distribution very often deviates slightly from the isotropic one, and the isotropic distribution is only the *most probable* one.

If we fix the end-to-end distance away from zero, the isotropic distribution is made impossible and the molecule assumes the distribution with most realizations given the constraint of fixed r. In other words, under this constraint we cannot reach the true unconstrained equilibrium. Rather a constrained equilibrium is reached with an entropy $S_E(r)$ which we shall presently calculate.

Analogous arguments can be formulated for the gas of N atoms. These are rather more commonly explained in the books than those for a rubber molecule so that we skip them. Suffice it to say that in the gas the distribution with most realizations is the homogenous one in the volume V, and that the gas has maximum entropy, if V is infinite. Whenever V is finite there is a pressure on the wall and that pressure is an entropic force.

We conclude that entropic forces may be expansive, like in a gas, or contractive, like in rubber. A circumstance where those tendencies compete and find an equilibrium is a gas-filled rubber balloon.

5.5 Entropy and disorder

It is often said, that the value of the entropy of a distribution is a measure for the disorder in the arrangement of its particles. This interpretation is once again most easily understood for the rubber molecule: Indeed, the stretched-out, orderly distribution has zero entropy because it can only be realized in one single manner. The disordered, kinky distribution has positive entropy. And the most disorderly isotropic distribution with very many possible realizations has the maximum value of entropy.

For a gas an analogous argument can be made: the order is perfect and the entropy is zero, when all atoms lie in *one* point and have *one* velocity. When more points are occupied, the order is less and the number of realizations is bigger and so is the entropy.

5.6 Molecular dynamics of the "entropic spring"

On the CD accompanying this book, in the program called "rubber", we exhibit a rubber molecule subject to thermal motion. The molecule is fixed at one end and may be loaded by a force on the other end. The temperature may be changed — hence the intensity of the thermal motion — and different forces may be applied. In this manner we demonstrate that the load-length-relation of the chain is linear like that of a linearly elastic spring. The rubber molecule is therefore sometimes called an *entropic spring*. Its mark of distinction is that the elastic modulus is proportional to the temperature.

5.7 Entropies of gases and of a rubber molecule

We proceed to calculate the equilibrium distributions N_{xc}^E and $N_{\vartheta\varphi}^E$, i.e. the distributions that maximize W in (5.11) and hence entropy. As constraints we adopt

$$\sum_{xc}^{PQ} N_{xc}=N, \quad \sum_{xc}^{PQ} \tfrac{\mu}{2}c^2 N_{xc}=U \quad \left\| \quad \sum_{\vartheta\varphi}^{PQ} N_{\vartheta\varphi}=N, \quad \sum_{\vartheta\varphi}^{PQ} b\cos\vartheta N_{\vartheta\varphi}=r, \quad (5.12)$$

where ϑ is the angle between a link and the axis through the end-points of the chain. The constraints will be taken care of by Lagrange multipliers. The procedure is standard and we obtain

$$N_{xc}^E = N\frac{e^{-\beta\frac{\mu}{2}c^2}}{\sum\limits_{xc}^{PQ} e^{-\beta\frac{\mu}{2}c^2}} \quad \left\| \quad N_{\vartheta\varphi}^E = N\frac{e^{-\beta b\cos\vartheta}}{\sum\limits_{\vartheta\varphi}^{PQ} e^{-\beta b\cos\vartheta}} \right. . \quad (5.13)$$

β is the Lagrange multiplier that takes care of the constraints concerning the internal energy U and the end-to-end distance r. It must be obtained from these constraints which — with (5.12) — may be written as

$$\frac{U}{N} = -\frac{\partial}{\partial\beta}\left(\ln\sum_{\vartheta\varphi}^{PQ} e^{-\beta\frac{\mu}{2}c^2}\right) \quad \left\| \quad \frac{r}{N} = -\frac{\partial}{\partial\beta}\left(\ln\sum_{\vartheta\varphi}^{PQ} e^{-\beta b\cos\vartheta}\right). \quad (5.14)$$

Insertion of (5.13) into (5.11) and the calculation of the equilibrium entropies $S^E = k\ln W_{\max}$ provides

$$S^E=Nk\left[\ln\left(\sum_{xc}^{PQ} e^{-\beta\frac{\mu}{2}c^2}\right)+\beta\frac{U}{N}\right]\left\|S^E=Nk\left[\ln\left(\sum_{\vartheta\varphi}^{PQ} e^{-\beta b\cos\vartheta}\right)+\beta\frac{r}{N}\right].\right.$$

$$(5.15)$$

Thus all that remains to be done is the calculation of the sums in (5.14) and (5.15). This is done by converting the sums into integrals.

In the gas the conversion follows the line of argument already explained in Sect. 3.6, by which the number of points $\mathbf{x},\mathbf{c}$ between $\mathbf{x}$, $\mathbf{c}$ and $\mathbf{x}+d\mathbf{x}$, $\mathbf{c}+d\mathbf{c}$ equals $XY\,d\mathbf{c}d\mathbf{x}$. Thus we obtain

$$\sum_{\mathbf{xc}} e^{-\beta\frac{\mu}{2}c^2} = XY \int\limits_{V} \int\limits_{-\infty}^{\infty} e^{-\beta\frac{\mu}{2}c^2}\,d\mathbf{c}d\mathbf{x} = XY\,V\sqrt{\frac{2\pi}{\beta\mu}}^{\,3}. \tag{5.16}$$

Thus from $(5.14)_1$ we obtain the Lagrange multiplier β in terms of U or temperature T

$$\beta = \tfrac{3}{2}\tfrac{1}{U/N} \quad\text{or by}\quad U/N = \tfrac{3}{2}kT \;:\; \beta = \tfrac{1}{kT}. \tag{5.17}$$

By $(5.13)_1$, $(5.15)_1$ the equilibrium distribution and the equilibrium entropy read with $\rho = \frac{N\mu}{V}$

$$N^E_{\mathbf{xc}} = \frac{1}{XY}\,\frac{\rho/\mu}{\sqrt{2\pi\frac{k}{\mu}T}^{\,3}}\,e^{-\frac{\mu c^2}{2kT}} \quad\text{and}\quad S^E = Nk\left[\ln\frac{T^{3/2}}{\rho} + \left\{\tfrac{3}{2} + \ln\left(XYN\mu\sqrt{2\pi\frac{k}{\mu}}^{\,3}\right)\right\}\right]. \tag{5.18}$$

Inspection shows that indeed the equilibrium distribution is homogeneous in space while the dependence on the velocity is given by the Maxwell distribution, cf. (3.17). The entropy $(5.18)_2$ conforms to the Boltzmann entropy in (3.38) at least concerning the T- and ρ-dependence. [We cannot expect the constants to agree, since the present calculation has added the term $k(N\ln N - N)$ to the Boltzmann entropy in (3.38). However, if that term is substracted from $(5.18)_2$ we obtain agreement between (3.38) and $(5.18)_2$.[3]]

For the rubber molecule we convert the sums in $(5.13)_2$, $(5.15)_2$ into integrals by assuming that the number of orientations ϑ, φ between ϑ, φ and $\vartheta+d\vartheta$, $\varphi+d\varphi$ is proportional to the solid angle element $\sin\vartheta\,d\vartheta\,d\varphi$ corresponding to that interval. If the factor of proportionality is Z, we obtain

$$\sum_{\vartheta\varphi}^{PQ} e^{-\beta b\cos\vartheta} = Z \int\limits_{0}^{2\pi}\int\limits_{0}^{\pi} e^{-\beta b\cos\vartheta}\sin\vartheta\,d\vartheta\,d\varphi = 4\pi Z\frac{\sinh\beta b}{\beta b} \tag{5.19}$$

and therefore, by $(5.14)_2$

$$\frac{r}{Nb} = -\left(\operatorname{ctgh}\beta b - \frac{1}{\beta b}\right). \tag{5.20}$$

The function in brackets is called the Langevin function; unfortunately it cannot be inverted analytically to give β as a function of $\frac{r}{Nb}$. However, an approximate inversion is possible, if we stick to cases where the molecule is strongly entangled, i.e. where $r \ll Nb$ holds. Since the Langevin function

---

[3] See Chap. 16 on the Gibbs paradox for a discussion of this point.

passes through the origin with the slope $1/3$, we have for the case of strong entanglement

$$\beta \approx -\frac{1}{3}\frac{r}{Nb^2}.$$

Thus the equilibrium distribution and the equilibrium entropy read

$$N_{\vartheta\varphi}^E = N\frac{1}{4\pi Z}\,e^{+\frac{r\cos\vartheta}{3Nb}} \quad \text{and} \quad S^E = Nk\left[\ln 4\pi Z - \frac{1}{3}\frac{r^2}{N^2b^2}\right]. \qquad (5.21)$$

We conclude that the equilibrium distribution, although independent of φ, is non-isotropic, unless $r = 0$ holds. The graph of the entropy is a concave parabola as a function of r so that the entropy is maximal for $r = 0$, cf. Sect. 5.4 and 5.5.

5.8 Entropy of a rubber rod

Modeling a rubber molecule as a chain of loose links is only the first step in modeling a rubber rod. For the rod we assume that it consists of a network of n such chains as shown in Fig. 5.6. The lines in that figure represent the chains, and the dots represent joints between them. The joints are points of entanglement of a molecule with itself or with neighbouring ones. Therefore a chain may be considerably shorter than a molecule. All chains are supposed to have N links, but their orientations are different and they are given by the cartesian components $(\vartheta_1, \vartheta_2, \vartheta_3)$ of their end-to-end distance vector. Thus the length r – or end-to-end distance – of a chain is $r = \sqrt{\vartheta_1^2 + \vartheta_2^2 + \vartheta_3^2}$. By (5.22) the entropy of a molecule is thus given by[4]

$$S_{\text{Mol}}(\vartheta_1, \vartheta_2, \vartheta_3) = Nk\left(\ln(4\pi Z) - \frac{\vartheta_1^2 + \vartheta_2^2 + \vartheta_3^2}{3N^2b^2}\right). \qquad (5.22)$$

The entropy of the rod is the sum of the chain entropies so that in the undistorted and in the deformed states of the rod we may write the entropies as

$$S_0 = \int S_{\text{Mol}} z_0(\vartheta_1, \vartheta_2, \vartheta_3)d\vartheta_1, d\vartheta_2, d\vartheta_3 \text{ and } S = \int S_{\text{Mol}} z(\vartheta_1, \vartheta_2, \vartheta_3)d\vartheta_1, d\vartheta_2, d\vartheta_3. \qquad (5.23)$$

$z_0 d\vartheta_1 d\vartheta_2 d\vartheta_3$ and $z d\vartheta_1 d\vartheta_2 d\vartheta_3$ are the numbers of distance vectors in the interval between $\vartheta_1, \vartheta_2, \vartheta_3$ and $\vartheta_1 + d\vartheta_1, \vartheta_2 + d\vartheta_2, \vartheta_3 + d\vartheta_3$ in the undisturbed and deformed states respectively.

---

[4] In this section we denote the equilibrium entropy of the molecules by S_{Mol} so as to distinguish it from the entropy of the rod which is denoted by S.

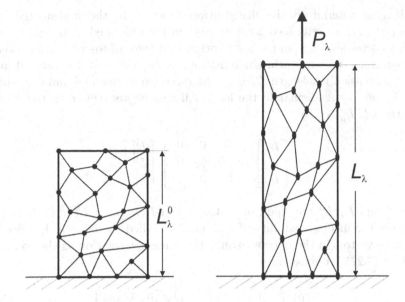

Fig. 5.6 A rubber rod in the undistorted state (left)
and deformed by the load P_λ (right).

We proceed to determine the distribution functions $z_0(\vartheta_1, \vartheta_2, \vartheta_3)$ and $z(\vartheta_1, \vartheta_2, \vartheta_3)$ and start with $z_0(\vartheta_1, \vartheta_2, \vartheta_3)$. Rubber in the undistorted state is isotropic so that we may assume that z_0 depends only on the distance r. If that distance is small, it can be realized in many ways, and if it is large it can be realized in few ways only. We assume that the number of distance vectors of length r in the network is proportional to the number of possibilities to realize r. And that number of realizations is given by

$$W = e^{\frac{1}{k} S_{\text{Mol}}} \tag{5.24}$$

according to the Boltzmann formula for entropy. Therefore we have by (5.21)

$$z_0\left(\vartheta_1, \vartheta_2, \vartheta_3\right) d\vartheta_1 d\vartheta_2 d\vartheta_3 = C\, e^{-\dfrac{\vartheta_1^2 + \vartheta_2^2 + \vartheta_3^2}{3N^2 b^2}}\, d\vartheta_1 d\vartheta_2 d\vartheta_3. \tag{5.25}$$

C is a factor of proportionality whose value may be calculated from the knowledge that the number of chains equals n. Thus

$$\int z_0 d\vartheta_1 d\vartheta_2 d\vartheta_3 = n, \text{ hence}^5\ C = \frac{n}{\sqrt{3\pi N b^2}^3}. \tag{5.26}$$

---

[5] The integration over ϑ_i $(i = 1, 2, 3)$ is carried from $-\infty$ to $+\infty$ without essential mistake.

Having determined the distribution function in the undistorted state $z_0(\vartheta_1, \vartheta_2, \vartheta_3)$ we still need $z(\vartheta_1, \vartheta_2, \vartheta_3)$ in the deformed state and we proceed to determine it from the assumption that the end-to-end distance vector of each chain deforms in the same manner as the rod itself. For the rod under uniaxial tension the length L_λ is in the direction of the load and the lengths L_μ, L_ν are perpendicular to the load. All lengths are related to the original lengths L_λ^0, L_μ^0, L_ν^0 by

$$\begin{bmatrix} L_\lambda \\ L_\mu \\ L_\nu \end{bmatrix} = \begin{bmatrix} \lambda & 0 & 0 \\ 0 & \frac{1}{\sqrt{\lambda}} & 0 \\ 0 & 0 & \frac{1}{\sqrt{\lambda}} \end{bmatrix} \begin{bmatrix} L_\lambda^0 \\ L_\mu^0 \\ L_\nu^0 \end{bmatrix}. \tag{5.27}$$

L_μ/L_μ^0 and L_ν/L_ν^0 are equal by isotropy and $L_\lambda L_\mu L_\nu = L_\lambda^0 L_\mu^0 L_\nu^0$ holds, because of the incompressibility of rubber; λ is called the stretch. If indeed all distance vectors in the network suffer the same deformation as the rod itself, namely (5.27), we must have

$$z(\vartheta_1, \vartheta_2, \vartheta_3) = z_0\left(\frac{1}{\lambda}\vartheta_1, \sqrt{\lambda}\vartheta_2, \sqrt{\lambda}\vartheta_3\right), \tag{5.28}$$

because a distance vector $\vartheta_1, \vartheta_2, \vartheta_3$ in the deformed state originally had the components $\frac{1}{\lambda}\vartheta_1, \sqrt{\lambda}\vartheta_2, \sqrt{\lambda}\vartheta_3$. With this relation between the functions z and z_0 and the knowledge of $z_0(\vartheta_1, \vartheta_2, \vartheta_3)$ from (5.25), (5.26) we conclude that

$$z(\vartheta_1, \vartheta_2, \vartheta_3) = \frac{n}{\sqrt{3\pi Nb^2}^3} e^{-\frac{\frac{1}{\lambda^2}\vartheta_1^2 + \lambda\left(\vartheta_2^2 + \vartheta_3^2\right)}{3Nb^2}}. \tag{5.29}$$

Insertion into (5.23) and simple integrations provide the entropy as a function of the stretch λ in the direction of the load. We obtain

$$S = nk\left(N\ln(4\pi Z) - \frac{1}{2}\left(\lambda^2 + \frac{2}{\lambda}\right)\right). \tag{5.30}$$

The graph of this function is shown on the left hand side of Fig. 5.7. It has a maximum in the undistorted state for $\lambda = 1$. For $\lambda \longrightarrow 0$ the entropy tends to $-\infty$ asymptotically and for $\lambda \longrightarrow \infty$ it approaches a concave parabola.

5.9 Uniaxial force-stretch curve

Having determined S as a function of the stretch λ we need only introduce that function into the equation (5.9) to obtain the force-stretch relation or the thermal equation of state for rubber. We get

$$P = \frac{nkT}{L_\lambda^0}\left(\lambda - \frac{1}{\lambda^2}\right). \tag{5.31}$$

The graph of that function is shown on the right hand side of Fig. 5.7. For $\lambda = 1$ we have $P = 0$, as it must be, and for $\lambda \longrightarrow 0$ we must apply a compressive force tending toward infinity. For large values of λ the force tends asymptotically to a linear function of λ.[6] The force is linear in T which must be expected; actually the proportionality of P and T was our starting point, cf. Fig. 5.3$_{\text{right}}$, when we identified the elasticity of rubber as entropic.

The thermal equation of state of rubber under uniaxial loading is the prototype of a thermal equation of state of a non-linear elastic body. It was first derived in the 1930's in a manner more or less as we have derived it here. In later times the state equation was found to be deficient but it still furnishes the leading term of the true force-stretch relation.[7]

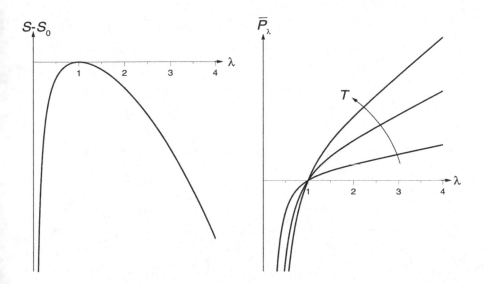

Fig. 5.7 Left: Entropy $S - S_0$ as a function of stretch λ.
Right: Force-stretch isotherms.

---

[6] The kinetic theory of rubber of the present simple form is restricted in validity to $\lambda \lesssim 3.5$, because for larger stretches crystallization occurs and the elasticity is no longer purely entropic.

[7] References and annotations may be found at the end of the book.

6

Statistical thermodynamics

While Boltzmann did discover the statistical interpretation of entropy, his arguments were pertinent and useful only for systems of independent elements like monatomic ideal gases, or rubber molecules, cf. Chap. 5. In real gases, or liquids, or solids the atoms interact and their statistical treatment requires the statistical thermodynamics of ensembles to which Boltzmann's ideas were extrapolated.

Statistical thermodynamics succeeds in expressing the thermodynamic equilibrium properties of arbitrary bodies in terms of a single function, *the partition function*. Most often the partition function cannot be calculated analytically, but it may sometimes be approximated. A case where it *can* be determined is the case of a hydrogen atom at rest in a heat bath.

6.1 On the need of statistical thermodynamics

The ease with which we have used

$$S = k \ln W \qquad \text{and} \qquad W = \frac{N!}{\prod\limits_{\mathbf{xc}}^{PQ} N_{\mathbf{xc}}!} \tag{6.1}$$

in Chap. 5 to calculate equations of state for ideal gases is deceptive. To be sure, the expressions (6.1) for entropy are always valid[1] as long as the state of a body can be characterized by the positions and velocities of its atoms. This is the case for many gases, vapours, liquids and solids. However, for all but the first case — gases — the energy is a complicated — at least quadratic and non-local — expression in $N_{\mathbf{xc}}$, because of the interatomic potential energy. Thus when we maximize S in (6.1) in order to find the equilibrium distribution $N_{\mathbf{xc}}^{E}$

---

[1] Well, let us pretend that this is so for the time being. We thus ignore the cautions issued in Sect. 3.6 and we refer to Chap. 16 on degeneracy and the pseudo-Gibbs-paradox for a qualification.

under the constraint of fixed energy, it turns out to be impossible to calculate N_{xc}^{E} analytically. Therefore we are stymied right at the beginning and there seems to be no way to proceed so as to obtain compact expressions for the equilibrium entropy, and the free energy, and equations of state.

This is where statistical thermodynamics comes in. Statistical thermodynamics succeeds in providing specific expressions for all thermodynamic equilibrium properties in terms of a single function, the *partition function*. And, while the partition function cannot very often   be determined analytically either as a function of thermodynamic variables — such as volume or deformation —, it may be obtained numerically or, in many cases, by reasonably plausible approximations.

We proceed to show how statistical thermodynamics works.

6.2 The canonical ensemble of identical bodies

We change focus and instead of regarding a body as an ensemble of N atoms we imagine a bigger body: An ensemble of ν subbodies of N atoms each. The energy of the ensemble is fixed and we denote it by ε. In each subbody the atoms are counted from 1 to N and each subbody's state is given by the positions and velocities of its atoms, i.e. $(\mathbf{x}_1...\mathbf{x}_N \ \mathbf{c}_1...\mathbf{c}_N)$. Among the ν subbodies let there be $\nu_{\mathbf{x}_1...\mathbf{c}_N}$ in the state $(\mathbf{x}_1...\mathbf{c}_N)$.

The energy of the subbody is given by the expression

$$E(\mathbf{x}_1...\mathbf{c}_N) = \sum_{\alpha=1}^{N} \frac{\mu}{2} c_\alpha^2 + \frac{1}{2} \sum_{\substack{\alpha,\beta=1 \\ \alpha \neq \beta}}^{N} \Phi(x_{\beta\alpha}) \qquad (6.2)$$

appropriate to a pairwise interaction — only within the subbody (!); the interaction potential Φ depends on the distance $x_{\beta\alpha} = |\mathbf{x}_\beta - \mathbf{x}_\alpha|$ of the interacting atoms. Therefore the energy of the ensemble is

$$\varepsilon = \sum_{\mathbf{x}_1...\mathbf{c}_N} E(\mathbf{x}_1...\mathbf{c}_N) \, \nu_{\mathbf{x}_1...\mathbf{c}_N}, \qquad (6.3)$$

where the summations extend over all positions and velocities available to the atoms of the subbodies.

In order to calculate the entropy σ of the ensemble we extrapolate (6.1) from a gas of N atoms to the ensemble of ν subbodies and write[2]

$$\sigma = k \ln W \qquad \text{with} \qquad W = \frac{\nu!}{\sum\limits_{\mathbf{x}_1...\mathbf{c}_N} \nu_{\mathbf{x}_1...\mathbf{c}_N}!} . \qquad (6.4)$$

---

[2] In this way the modification (3.36) of Boltzmann's formula (3.35) was canonized in statistical thermodynamics and the road to the pseudo-Gibbs-paradox was paved.

The equilibrium distribution $\nu_{x_1 \ldots c_N}$ is calculated as the one that maximizes σ under the constraint of constant ε — and of constant ν, of course. We obtain what is called a *canonical distribution*

$$\nu_{x_1 \ldots c_N}^E = \nu \frac{e^{-\beta E(x_1 \ldots c_N)}}{\sum\limits_{x_1 \ldots c_N} e^{-\beta E(x_1 \ldots c_N)}}. \qquad (6.5)$$

Hence follows for the ensemble

$$\varepsilon = -\nu \frac{\partial \ln P}{\partial \beta} \quad \text{where} \quad P = \sum_{x_1 \ldots c_N} e^{-\beta E(x_1 \ldots c_N)} \sim \text{partition function}$$

$$\sigma = \nu k \left(\beta \frac{\varepsilon}{\nu} + \ln P \right). \qquad (6.6)$$

β is the Lagrange multiplier that takes care of the energy constraint. Because of (6.2) the kinetic energy of the ensemble is given by

$$\varepsilon_{\text{kin}}^E = \sum_{x_1 \ldots c_N} \left(\frac{\mu}{2} c_1^2 + \ldots \frac{\mu}{2} c_N^2 \right) \nu_{x_1 \ldots c_N}^E = \nu N \frac{3}{2} \frac{1}{\beta}. \qquad (6.7)$$

Since we expect that the mean kinetic energy of an atom is equal to $\frac{3}{2} kT$ [3], we conclude that $\beta = \frac{1}{kT}$ holds.

The mean energy and entropy of a subsystem is therefore

$$U = \frac{\varepsilon}{\nu} = kT^2 \frac{\partial \ln P}{\partial T} \quad \text{where} \quad P = \sum_{x_1 \ldots c_N} e^{-\frac{E(x_1 \ldots c_N)}{kT}}$$

$$S = \frac{\sigma}{\nu} = k \frac{\partial}{\partial T} (T \ln P), \quad \text{hence the free energy} \quad F = U - TS \qquad (6.8)$$

$$F = -kT \ln P.$$

Thus we have done what we proposed to do in Sect. 6.1: We have reduced the thermodynamic functions of a system of interacting atoms to the calculation of the partition function. What we have described is a neat way to bypass the difficulties of exploiting the maximum of the entropy (6.1) for such a complex body, cf. Sect. 6.1. What remains is the difficulty of evaluating the partition function. Later, in Chapt. 13, when we deal with shape memory alloys, we shall encounter a non-trivial example.

6.3 Discussion of the canonical ensemble. Ergodic hypothesis

The ensemble of ν subbodies with fixed energy is called a canonical ensemble in statistical thermodynamics. Its introduction was a stroke of genius by Gibbs

---

[3] Statistical mechanics proves this but we skip that argument.

by which he pushed aside scruples that might have stymied lesser minds.[4] Let us consider:

To fix the ideas we may think of the subbodies as being in individual containers. The atoms in different containers do not interact by intermolecular forces and yet they are in "thermal contact" so that in equilibrium they all have the same temperature T.

Of course, the *composite body* exists only in imagination and therefore it might seem to be irrelevant to know that the number $\nu^E_{\mathbf{x}_1...\mathbf{c}_N}$ of its subbodies are given by (6.5) in equilibrium. However, there is more to this than we have hitherto discussed: Indeed, the *ergodic hypothesis* assumes that $\nu^E_{\mathbf{x}_1...\mathbf{c}_N}$ is also the frequency of the state $\mathbf{x}_1...\mathbf{c}_N$ in a *single subbody* of the equilibrated ensemble, if that subbody is obvserved ν times at sufficiently large intervals. This hypothesis is eminently plausible and generally accepted.[5] Therefore U, and S, and F in (6.8) may be considered as average equilibrium values of energy, entropy and free energy of a single subbody of temperature T – averaged over many consecutive observations of that subbody.

Ultimately one forgets about the ensemble altogether and speaks of U, S and F in (6.8) as energy, entropy and free energy of a body in equilibrium with a *heat bath* of temperature T.

6.4 Further extrapolation. Entropy of a hydrogen atom

So far we have spoken of systems of atoms whose states are characterized by a position and a velocity. But statistical thermodynamics has a much wider scope. It is applied to any system that participates in the ubiquitous thermal motion.

As an extreme example we proceed to calculate the entropy of an H-atom at rest. We trust that the reader possesses the knowledge of the Bohr model of atomic structure and knows that an H-atom has a single electron which may occupy $2n^2$ orbits ($n = 1, 2, ..$) of energies[6]

$$E_n = \frac{2\pi^2 e^4 m}{(4\pi\varepsilon_0 h)^2}\left(1 - \frac{1}{n^2}\right) = 2.171 \cdot 10^{-18}\text{J}\left(1 - \frac{1}{n^2}\right) \qquad (6.9)$$

[4] See also "References and annotations" at the end of the book.

[5] A trivial example is this: Suppose that on an aerial photo of a city you identify the fraction of cars which drive with 50 km/h. Next, consider that you drive a car yourself for some long time in the city and calculate the fraction of seconds that your speed is 50 km/h. The ergodic hypothesis says that the two fractions are equal. Mathematicians have tried to prove the ergodic hypothesis and their efforts have led to a branch of set theory, the ergodic theory. That theory, however, offers little to the physicist.

[6] Thus the energy of the atom is set equal to zero when the electron is in the ground state n=1. [We may use any other convention for that energy; all choices will lead to the same entropy.]

where e and m are the electric charge and mass of the electon and $h = 6.625 \cdot 10^{-34}$ J sec is the Planck constant: ε_0 is the vacuum dielectric constant. The factor 2 in $2n^2$ comes from the fact that the electron may occupy an orbit with either spin up or spin down. Both states have the same energy.

In the jargon of the previous section we place the atom into a heat bath and use (6.8) to calculate its partition function as

$$P = \sum_{n=1}^{\infty} 2n^2 \, e^{-\frac{E_n}{kT}}. \tag{6.10}$$

Therefore the entropy of the H-atom in equilibrium with the heat bath is equal to

$$S_H = k \frac{\partial}{\partial T} \left(T \ln \left[\sum_{n=1}^{\infty} 2n^2 \, e^{-\frac{E_n}{kT}} \right] \right). \tag{6.11}$$

For any "normal" earthly temperature only the first term in (6.10) contributes significantly to the partition function and therefore the entropy of an H-atom may be calculated as

$$S_H = k \ln 2 \qquad \forall \ \ T < 10^4 K. \tag{6.12}$$

Later in Chapt. 14 we shall use this value to calculate the entropy constant of an atomic hydrogen gas.

7

Entropy and energy in competition

Equivalent to the growth of entropy in an adiabatic body is the decrease of
the available free energy in a body whose boundary is kept at a constant
temperature. The form of the available free energy depends on the nature
of the working on the body. The working of conservative body forces may
be represented by a potential energy but the working on the boundary of
the body may assume quite a variety of forms. Special cases are
- no working, when the surface of the body is at rest
- working by a dead load.

Generally the available free energy contains an energetic part and an entropic part. And when the free energy assumes a minimum, we may consider
this as a compromise between the energy which tends to a minimum and the
entropy which tends to a maximum. Those two tendencies — the energetic
and the entropic one — often compete, since generally the energy favours a
different distribution of matter than the entropy.

7.1 A succinct formulation of the thermodynamic laws

Clausius and Gibbs were first in having a full grasp of thermodynamics and
they summarized the 1st and 2nd laws in the triumphant slogan

Die Energie des Weltalls ist konstant.
Die Entropie des Weltalls strebt einem Maximum zu.

The significance of choosing the universe as the exemplary body lies in the
fact that we may be sure, perhaps, that the universe has an adiabatic surface
and that no working is done on its surface.

Clausius describes the final equilibrium with a maximal entropy dramati-
cally as the *heat death* of the universe. We shall attempt to follow his line of
thought and illustrate the idea for a simple model.

First of all, however, let us realize that the universe is replete with bodies
that attract each other gravitationally and that a large part of the energy of
the universe resides in the potential energy of gravitation. Therefore we proceed to introduce the gravitational potential energy into the 1st law explicitly,

7.2 Potential energies

Actually gravitational forces occur in the 1st law (1.1) already, because the working $\dot{A}$ on a body in the volume V has two parts, an external one due to stress on the surface and an internal one due to body forces, in particular gravitational forces:

- The working of the stress t_{ij} on the elements dA of the surface ∂V reads

$$\dot{A}_{ext} = \int_{\partial V} t_{nj} v_n n_j dA, \tag{7.1}$$

  where v_n are the components of velocity and n_j is the outer unit normal on ∂V.
- The working of the specific body force f_n on the mass elements dm inside V reads

$$\dot{A}_{int} = \int_V f_n v_n dm, \quad \text{or with} \quad f_n = -\frac{\partial \varphi}{\partial x_n} \tag{7.2}$$

$$= \frac{d}{dt}\left(-\int_V \varphi dm\right) \quad \text{and with} \quad E_{Pot} = \int_V \varphi dm \tag{7.3}$$

$$\dot{A}_{int} = -\frac{dE_{Pot}}{dt}. \tag{7.4}$$

The relation $(7.2)_2$ holds for *conservative* forces, e.g. gravitational forces, and φ in that case is the gravitational potential.[1]

The 1st law (1.1) may thus be written in the form

$$\frac{d\left(U + E_{Pot} + K\right)}{dt} = \dot{Q} + \dot{A}_{ext}. \tag{7.5}$$

We note that E_{Pot}, defined by $(7.3)_2$, represents the potential energy of external forces, e.g. the gravitational force, or inertial forces, or electro-magnetic forces. This is not the only occurence of potential energy in the 1st law (7.5), however. Indeed, U itself contains − in its V-dependence − the potential energy of the interatomic forces. In the simple case treated by van der Waals the effect of internal potential energy is made explicit, cf. Fig. 4.2.

---

[1] For $\dot{A}_{int}$ to have the form (7.4) it is usually necessary that φ be independent of t. An exception is self-gravitation of a star, where φ depends on $\mathbf{x}$ and on t in a particular manner.

7.3 Growth of entropy under the constraint of constant energy

The most immediate consequence of the 2nd law $(1.24)_2$ for a rapid or irreversible process is that the entropy grows in an adiabatically isolated body. Thus, while equilibrium is approached, the entropy $S(t)$ tends to a maximum.

To be sure, during the irreversible process, the fields of temperature $T(\mathbf{x}, t)$ and density $\rho(\mathbf{x}, t)$ – or specific volume $v(\mathbf{x}, t) = 1/\rho(\mathbf{x}, t)$– may be strongly time-dependent and non-homogeneous in the body and there may be considerable kinetic energy in the body due to a flow field with velocity $\mathbf{v}(\mathbf{x}, t)$. Therefore the entropy $S(t)$ must be written as an integral over the specific entropies $s(\mathbf{x}, t)$ of the mass elements dm. We thus require[2]

$$S(t) = \int_V s(T(\mathbf{x}, t), v(\mathbf{x}, t))dm \;\rightarrow\; \text{maximum}, \qquad (7.6)$$

where $s(\mathbf{x}, t)$ as indicated in (7.6), is most often considered to be a function of $T(\mathbf{x}, t)$ and $v(\mathbf{x}, t)$ – the "local state" of dm at point $\mathbf{x}$ and time t – such that the Gibbs equation (1.19) holds locally and at all times for each mass element[3]

$$ds = \frac{1}{T}(du + pdv). \qquad (7.7)$$

The maximum of $S(t)$ is not approached in an unconstrained manner, since the 1st law (7.5) must be obeyed during the process. In the special case, when there is neither heating nor external working, this constraint reads

$$U + E_{Pot} + K = \int_V \left(u(T(\mathbf{x}, t), v(\mathbf{x}, t)) + \varphi(\mathbf{x}, t) + \frac{1}{2}v^2(\mathbf{x}, t) \right) dm = \text{constant},$$
$$(7.8)$$

where $u(\mathbf{x}, t)$ and $\frac{1}{2}v^2(\mathbf{x}, t)$ are the fields of the specific values of internal and kinetic energy.

7.4 An example: Piston drops into a cylinder

We consider the adiabatically isolated body shown in Fig. 7.1 consisting of a piston fixed on top of a gas-filled cylinder of cross-sectional area A. Initial conditions before the piston is released are given by

---

[2] We follow the usual practice and denote global quantities by capital letters, and specific quantities by the corresponding minuscules.

[3] This assumption is known as the principle of local equilibrium. In order to appreciate that this is indeed an assumption, one should realize that $s(\mathbf{x}, t)$ might also depend on the gradients and/or rates of change of $T(\mathbf{x}, t)$ and $v(\mathbf{x}, t)$, when those fields are non-homogeneous and time-dependent.

$$T(\mathbf{x}, t^i) = T^i \text{ homogeneous, } H^i, \text{ and } \mathbf{v}(\mathbf{x}, t^i) = 0 \qquad (7.9)$$

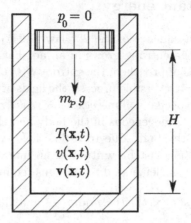

Fig. 7.1 A body consisting of piston and a gas-filled cylinder.

When the piston is released, it will move either downwards or upwards and, while it moves, it creates a flow in the gas – usually a turbulent flow. The motion of the piston will also create non-homogeneous fields of temperature, density and pressure. We ask for the final fields $T(\mathbf{x}, t^f)$, $p(\mathbf{x}, t^f)$ and $\mathbf{v}(\mathbf{x}, t^f)$ of temperature, pressure and velocity and for the final height H^f of the piston when equilibrium has been established so that the entropy is maximal.

While the engineer sees a part of a shock absorber in this example, the physicist – with some stretch of imagination – sees a model for the universe, perhaps. We shall try to accommodate both points of view.

First of all we take the problem at face value and solve it. For simplicity we consider the mass m_P of the piston as concentrated at the height H so that, with the gravitational potential $\varphi = gx_3$, the potential energy of the piston is equal to[4]

$$E_{Pot} = \int\limits_V \varphi(\mathbf{x}) \, dm = m_P g H = \frac{m_P g}{A} \int\limits_V v(\mathbf{x}, t) \, dm , \qquad (7.10)$$

where g is the gravitational acceleration. We denote the mass of the gas by m_G and neglect its potential energy. And we assume that there is no external working so that the constraint (7.8) applies. We maximize S in (7.6) under that constraint, which is taken care of by the Lagrange multiplier λ. Thus we must maximize the functional

---

[4] The integral in $(7.10)_2$ is a complicated but useful way of writing V. It is useful, because $v(\mathbf{x}, t)$ at each point $\mathbf{x}$ are variables in the subsequent maximization problem.

$$\int_V s\left(T\left(\mathbf{x},t\right),v\left(\mathbf{x},t\right)\right)dm -$$

$$\lambda\left\{\int_V\left(\left(u(T\left(\mathbf{x},t\right),v\left(\mathbf{x},t\right)\right)+\frac{m_P g}{A}v\left(\mathbf{x},t\right)+\frac{1}{2}v^2\left(\mathbf{x},t\right)\right)dm\right\}.$$

The independent variables are the values of 5 fields $T(\mathbf{x},t)$, $v(\mathbf{x},t)$ and $v_n(\mathbf{x},t)$ in all interior points and the necessary conditions for a maximum read

$$\frac{\partial s}{\partial T}=\lambda\,\frac{\partial u}{\partial T}\qquad\qquad\qquad\frac{1}{T\left(\mathbf{x},t\right)}=\lambda$$

$$\frac{\partial s}{\partial v}=\lambda\left(\frac{\partial u}{\partial v}+\frac{m_P g}{A}\right)\ \text{hence by (7.7)}\ \ p(\mathbf{x},t)\ =\frac{m_P g}{A}\qquad(7.11)$$

$$v_n\ =0\qquad\qquad\qquad\qquad v_n(\mathbf{x},t)\ =0\,.$$

It follows that the body is homogeneously at rest in equilibrium and that the temperature field is homogeneous. So is the pressure field, and its value is equal to the pressure of the piston.

We imagine for simplicity that the piston has an adiabatically isolated bottom so that its temperature does not change. Thus, by use of (7.8) and (7.11), we obtain for the gas

$$T^f=\frac{z}{z+1}T^i+\frac{1}{z+1}\frac{m_P g}{m_G\,k/\mu}H^i,\quad H^f=\frac{1}{z+1}H^i+\frac{z}{z+1}\frac{m_G\,k/\mu}{m_P g}T^i\,.$$
$$(7.12)$$

The temperature has risen, if the piston has dropped and vice versa, i.e.

$$T^f\gtrless T^i\quad\text{if}\quad H^f\lessgtr H^i\,.\qquad\qquad(7.13)$$

Inspection of (7.12) shows that the ratio m_P/m_G is the essential parameter which determines which of the two possibilities will prevail.

It is instructive to discuss the limiting case of an infinitely massive piston. In that case we have

$$T^f_\infty=\infty\qquad\text{and}\qquad H^f_\infty=\tfrac{1}{z+1}H^i\qquad\qquad(7.14)$$

so that the piston − irrespective of its weight − cannot compress the gas to zero volume. Indeed, the pressure p^f tends to infinity along with T^f at a finite volume and that keeps the infinitely heavy mass in equilibrium.

So much about the shock-absorber-aspect of the problem. The imaginative physicist, for whom Fig. 7.1 models the universe, will consider this: Before equilibrium is reached, a hypothetical population living in the cylinder has quite some energy sources *available* to it for conversion into useful work: The dropping piston, pressure gradients inside the gas, temperature gradients, and a lively flow of the gas. When equilibrium is reached, however, the piston comes to rest, the gradients of pressure and temperature die out, and so does the flow of the gas. There is nothing to exploit anymore, the environment has become stale — and hot! That is what Clausius means by "heat death" as the final destiny of the universe.

This works also in the opposite direction. If the parameter m_P/m_G — and the initial conditions — are such that the piston rises after being released, it will still create a complicated flow in the cylinder and strongly non-homogeneous fields of density and temperature which are *available* for the creation of useful work. The entropy grows in that case as well; it grows until a maximum is reached and the system becomes again stale — but now cold! One is tempted to paraphrase Clausius in this expansive process and call the final state the "cold death".

7.5 Available free energies

The key word in the preceding discussion is "available": As long as equilibrium is approached, but not reached yet, energy is available inside the cylinder for conversion into useful work. Since the body is adiabatically isolated we may say that it is the shortfall of entropy from its maximum that determines the availability of energy; when the entropy approaches its maximum, less and less energy is available. It is also said that the increase of entropy "drives" the approach to equilibrium. We proceed to extend this idea to non-adiabatic bodies and to bodies with either surfaces at rest or subject to external working.

If the body has a non-adiabatic surface and if it is subject to external working, it will generally not tend to an equilibrium, because the interference from external heating and working prevents that approach.

There are exceptions, however, and they all concern the case in which the heating occurs through a surface which has the constant and homogeneous temperature T_0. In that case the 1st and 2nd laws may be combined — by elimination of $\dot{Q}$ — to give, cf. (7.5) and (1.24)$_2$

$$\frac{\mathrm{d}\,(U + E_{\text{Pot}} + K - T_0 S)}{\mathrm{d}t} \leq \dot{A}_{\text{ext}} \overset{!}{\underset{\text{cf. (7.1)}}{=}} \int_{\partial V} t_{nj} v_n n_j \, \mathrm{d}A. \qquad (7.15)$$

Within this case of constant and homogeneous boundary temperature T_0 we distinguish situations that differ by a specific form of $\dot{A}_{\text{ext}}$ and for which we expect equilibrium to prevail eventually.

• The first such case concerns surfaces at rest, so that $\dot{A}_{\text{ext}} = 0$ holds. In this situation (7.15) implies

$$\frac{\mathrm{d}\,(U + E_{\text{Pot}} + K - T_0 S)}{\mathrm{d}t} \leq 0 \qquad (7.16)$$

and we conclude that it is the decrease of $U + E_{\text{Pot}} + K - T_0 S$ which "drives" the approach to equilibrium.

- Alternatively, if there is external working but it is exerted by a constant and homogeneous pressure p_0 on the surface ∂V — or on part of it, while the rest of ∂V is either stress-free or at rest — we have from (7.1), or $(7.15)_2$

$$\dot{A}_{\text{ext}} = -p_0 \frac{\mathrm{d}V}{\mathrm{d}t} \ . \tag{7.17}$$

In that case $(7.15)_1$ implies

$$\frac{\mathrm{d}\left(U + E_{\text{Pot}} + K - T_0 S + p_0 V\right)}{\mathrm{d}t} \leq 0 \ , \tag{7.18}$$

and it is the decrease of $U + E_{\text{Pot}} - T_0 S + p_0 V$ which "drives" the approach to equilibrium.

This consideration leads us to define *available free energies*, generically denoted by $\mathfrak{A}$, as the quantities which tend to a minimum as equilibrium is approached. If $\mathfrak{A}$ exceeds its minimum, energy is available (sic!) inside the body for conversion into useful work.

Alternative forms of the available free energy, other than

$$\mathfrak{A} = \begin{cases} U + E_{\text{Pot}} + K - T_0 S & \mathbf{v} = 0 \ \text{on} \ \partial V \\ & \text{for} \\ U + E_{\text{Pot}} + K - T_0 S + p_0 V & \begin{array}{l} p = p_0 \sim \text{const} \ \& \\ \text{homogeneous on} \ \partial V \end{array} \end{cases} \tag{7.19}$$

result from different forms of the external working $\dot{A}_{\text{ext}}$. The stress in the working integral may be given by different tensile stress components, or by shear stresses, etc. In such cases the term $p_0 V$ in $(7.19)_2$ must be replaced by an expression appropriate to the loading situation. We do not know the general form of that expression and probably no such general form exists. The available free energy $\mathfrak{A}$ must therefore be determined from case to case.

7.6 Minimization of available free energies

The availabilities are functionals of the temperature field $T(\mathbf{x}, t)$, the density field $\rho(\mathbf{x}, t)$, and the velocity field $\mathbf{v}(\mathbf{x}, t)$ of the flow inside ∂V. Therefore we may write (7.19) in the more explicit form[5]

$$\mathfrak{A} = \int\limits_V \left[\rho(\mathbf{x}) \left(u\left(T(\mathbf{x}), \rho(\mathbf{x})\right) + \varphi(\mathbf{x}) + \frac{1}{2} v^2(\mathbf{x}) - T_0 s\left(T(\mathbf{x}), \rho(\mathbf{x})\right) \right) + \{p_0\} \right] \mathrm{d}V. \tag{7.20}$$

---

[5] For brevity we drop the variable t.

The term in curly brackets — the one with p_0 — belongs to $(7.19)_2$; it is absent in the case $(7.19)_1$. The variables are the values of the 5 fields $T(\mathbf{x})$, $v(\mathbf{x}) = 1/\rho(\mathbf{x})$ and $v_n(\mathbf{x})$ in the interior points of V. There is a constraint on them, because the total mass is constant

$$m = \int_V \rho(\mathbf{x})\mathrm{d}V \ . \qquad (7.21)$$

We minimize (7.20) and take care of the constraint (7.21) by a Lagrange multiplier λ. Thus, by use of the local Gibbs quation (7.7) we obtain as equilibrium conditions

$$v_n(\mathbf{x}) = 0$$

$$T(\mathbf{x}) = T_0$$

$$\qquad (7.22)$$

$$\underbrace{u(T_0, \rho(\mathbf{x})) - T_0 s(T_0, \rho(\mathbf{x})) + \frac{p(T_0, \rho(\mathbf{x}))}{\rho(\mathbf{x})}}_{g(T_0, \rho(\mathbf{x}))} + \varphi(\mathbf{x}) - \lambda = 0 \rightarrow \rho = \rho(\mathbf{x}, T_0, \lambda)$$

The first two of these conditions are self-explanatory. They imply that in equilibrium the body is at rest and has a homogeneous temperature field. The last condition $(7.22)_3$ may be used — as indicated — to calculate the density field in equilibrium in terms of the parameter T_0 and the Lagrange multiplier λ. The Lagrange multiplier must then be calculated from the constraint (7.21).

The quantity in braces in $(7.22)_3$ is variably called the specific free enthalpy, or the specific Gibbs free energy, or the chemical potential. The latter expression is unfortunate, since the whole argument has nothing to do with chemistry — except in the most far-fetched sense. But, however unfortunate, in the present instance the expression allows us to formulate $(7.22)_3$ neatly by the words: The sum of the chemical and gravitational potentials is homogeneous in equilibrium.

An explicit calculation of $\rho(\mathbf{x})$ from $(7.22)_3$ requires the knowledge of the equations of state $p(T, \rho)$, $u(T, \rho)$ and $s(T, \rho)$ — or $g(T, \rho)$. An instructive example is presented in the next chapter, where we shall consider planetary atmospheres.

A remark on the kinetic energy K: The kinetic energy of the flow field inside the body is part of the available free energy and an important one, since it can be converted into useful work. However, in equilibrium the body is invariably at rest — see (7.11) and (7.22) for instance — and K assumes its minimum, namely zero. In fact, the minimization of K does not affect the result of the minimization of the rest of $\mathfrak{A}$, as long as we recall that $v=0$ holds in equilibrium. Therefore, for the purpose of the calculation of equilibrium properties, we may just as well drop K from the available free energies. And we shall do so in the sequel.

7.7 Energy and entropy as competitors

Leaving the kinetic energy aside, as explained, and concentrating on a case with $T(\mathbf{x}) = T_0 = \text{const}$, we see that the approach to equilibrium consists of a rearrangement of the mass distribution in V. In a manner of speaking this rearrangement is driven by the tendency of the available free energy[6]

$$\mathfrak{A} = U + E_{\text{Pot}} + \{p_0 V\} - T_0 S \qquad (7.23)$$

to approach a minimum.

Inspection identifies two situations that are conducive to a minimum of $\mathfrak{A}$:
- A minimum of energy $U + E_{\text{Pot}} + \{p_0 V\}$ and
- a maximum of entropy.

Generally the energetic minimum and the entropic maximum favour different mass distributions. For a small temperature T_0, where we may ignore the term $T_0 S$, the mass distribution will be such that it minimizes the energy. On the other hand, for a large temperature T_0, the term $T_0 S$ in (7.23) prevails and the mass will arrange itself so as to maximize entropy.

Thus there is a competition between energy and entropy and as a result of that competition neither the energy attains a minimum nor the entropy a maximum. Rather the available free energy attains a minimum. That resultant minimum is close to the energetic one for a low temperature and close to the entropic maximum for a high temperature.

Indeed, much of thermodynamics may be seen in this light, and subsequent chapters contain several examples for the competition of energy and entropy. We begin with a consideration of the atmosphere of planets.

---

[6] The term $p_0 V$ is present or not depending on which one of the two cases in (7.19) prevails.

8

Planetary atmospheres

The atmosphere of a planet is a case in point for the competition of energy and entropy. Energy tends to assemble all atmospheric molecules on the surface of the planet while entropy tends to spread the molecules evenly throughout space. In the end the entropy wins and planets will be bare of atmosphere. This will happen earlier for a hot planet than for a cold one. Earth, which is a temperate planet in the solar system, has already lost all light gases but it holds on — for the time being — to the heavier ones like nitrogen, oxygen and argon.

8.1 Model atmosphere

We consider an atmosphere of total mass M_A of a planet as an ideal gas of constant and uniform temperature $T(\mathbf{x}) = T_0 = \text{const}$. Therefore the internal energy U is constant. The density distribution in the atmosphere will arrange itself so as to minimize the available free energy which in this case has only two essential parts: The potential energy and the entropy. The potential energy will tend to assemble the atoms of the atmosphere on the surface of the planet, because that is where the potential energy has its minimum. On the other hand, the entropy will tend to become maximal by scattering the atoms through infinite space. Thus the planetary atmosphere is subject to opposing tendencies and it provides a paradigm for the competition of energy and entropy.

In order to investigate this competition suggestively we convert the available free energy from a functional of $\rho(\mathbf{x})$ to a function of a single variable by considering a model atmosphere which is constrained between the planetary surface and an imagined roof of fixed height H. We consider this model atmosphere as being in *partial* equilibrium in the sense that the available free energy under the roof is minimal. In this manner $\mathfrak{A}$ — and E_{Pot} and S — become functions of the single variable H and we investigate their extrema in order to determine the density distribution in *full unconstrained* equilibrium, and in order to see what the roles of E_{Pot} and S are in establishing that equilibrium.

Since the height H is fixed, the available free energy reads

$$\mathfrak{A} = U + E_{\text{Pot}} - T_0 S. \tag{8.1}$$

U is constant in this case so that it does not affect the minimization problem, nor does the constant and purely T-dependent part of S. Therefore the essential part of $\mathfrak{A}$, denoted by $\tilde{\mathfrak{A}}$, has the form

$$\tilde{\mathfrak{A}} = \int_V \rho(\mathbf{x}) \left\{ \varphi(\mathbf{x}) + \frac{k}{\mu} T_0 \ln \frac{\rho(\mathbf{x})}{\rho(T_R, p_R)} \right\} d\mathbf{x}. \tag{8.2}$$

8.2 Thin-atmosphere-approximation

For a first calculation we assume that the atmosphere is close enough to the surface of the planet, so that everywhere in the atmosphere the gravitational potential φ equals $\varphi = gz$, where g is the gravitational acceleration on the surface. We consider a cylinder of cross-section A and height H, so that $d\mathbf{x} = A dz$ holds in (8.2). Also we assume that $\rho(\mathbf{x})$ depends only on the vertical component z. Therefore we have

$$\tilde{\mathfrak{A}} = A \int_0^H \rho(z) \left\{ gz + \frac{k}{\mu} T_0 \ln \frac{\rho(z)}{\rho(T_R, p_R)} \right\} dz. \tag{8.3}$$

We determine the density distribution $\rho(z)$ that minimizes $\tilde{\mathfrak{A}}$ under the constraint of constant mass

$$M_A = A \int_0^H \rho(z) dz. \tag{8.4}$$

An easy calculation provides[1]

$$\rho(z; H) = \frac{M_A}{AR} \beta \frac{e^{-\beta x}}{1 - e^{-\beta h}}, \quad \text{where} \quad \beta = \frac{gR}{k/\mu T_0}, \; x = \frac{z}{R}, \; h = \frac{H}{R}. \tag{8.5}$$

The mean height of the mass distribution results as

$$\bar{z}(H) = R \frac{1}{\beta} \left(1 - \frac{\beta h e^{-\beta h}}{1 - e^{-\beta h}} \right) \rightarrow \begin{cases} 0 & H \to 0 \\ & \text{for} \\ \dfrac{k/\mu T_0}{g} & H \to \infty \end{cases} ; \tag{8.6}$$

the latter term is obviously the mean height of the unconstrained isothermal atmosphere as given by the barometric formula.

---

[1] R is an arbitrary length used for the non-dimensionalization in (8.5). In order to be specific we take it to be equal to the radius of the planet.

We calculate $E_{\text{Pot}}(H)$ and the part $\tilde{S}(H)$ of S that contributes to $\tilde{\mathfrak{A}}$ in (8.2)

$$E_{\text{Pot}}(H) = M_A \frac{k}{\mu} T_0 \left(1 - \frac{\beta h e^{-\beta h}}{1 - e^{-\beta h}} \right)$$

(8.7)

$$\tilde{S}(H) = \frac{E_{\text{Pot}}}{T_0} + M_A \frac{k}{\mu} \ln \left[\frac{1}{\beta} \left(1 - e^{-\beta h} \right) \right] + \text{const.}$$

The constant in $\tilde{S}$ is unimportant, since it does not depend on H. The essential part of the available free energy $\tilde{\mathfrak{A}}$ thus reads

$$\tilde{\mathfrak{A}}(H) = -M_A \frac{k}{\mu} T_0 \ln \left[\frac{1}{\beta} \left(1 - e^{-\beta h} \right) \right] - T_0 \cdot \text{const.}$$

(8.8)

Figure 8.1 shows plots of these functions which we proceed to interpret. To within sign the slope of $\tilde{\mathfrak{A}}(H)$ determines the force P which the atmosphere exerts on the roof. Indeed, by (8.8), (8.5) we have

$$\frac{d\tilde{\mathfrak{A}}}{dH} = -A\, \rho(H; H) \frac{k}{\mu} T_0, \quad \text{hence} \quad \frac{d\tilde{\mathfrak{A}}}{dH} = -P.$$

(8.9)

We may say, therefore, that the slope of $\tilde{\mathfrak{A}}$ indicates the intensity with which the gas attempts to move the roof in its desire to reach an unconstrained equilibrium. And since the slope of $\tilde{\mathfrak{A}}$ is consistently negative, the roof will be pushed to infinity, albeit with a smaller and smaller force, because $\tilde{\mathfrak{A}}$ – according to Fig. 8.1 – drops asymptotically to a constant value. This, however, does not mean that the atmosphere is scattered into infinite space; indeed, $\bar{z}(H)$ is finite for all values H, including $H \to \infty$, cf. (8.6) and Fig. 8.1. Thus we conclude – at least from this calculation – that the atmosphere should remain with the planet.

Let us consider another aspect of the force P: Since $\tilde{\mathfrak{A}}$ equals $E_{\text{Pot}} - T_0 \tilde{S}$ we may write

$$P = -\frac{dE_{\text{Pot}}}{dH} + T_0 \frac{d\tilde{S}}{dH}$$

(8.10)

and we may interpret this equation by saying that the force on the roof has two parts, an energetic one and an entropic one.

According to Fig. 8.1 E_{Pot} is a monotonically increasing function of H. Therefore the energetic part of P is negative, or contractive, and would suck the roof down. That is perfectly understandable. Indeed, if it were for energy alone, the atoms of the atmosphere would all lie at the bottom at $H = 0$ – along with the rocks and the water – where E_{Pot} has its minimum.

The entropic part of the force, however, is positive since $\tilde{S}$ – according to Fig. 8.1 – grows monotonically. Therefore the entropic force is expansive. In fact the entropic force is infinitely strong for $H = 0$ — because of the infinite slope of the $\tilde{S}(H)$ graph — and is thus easily able to counteract the energetic tendency to assemble the atoms at the bottom. For larger values of H the entropic force becomes weaker and weaker.

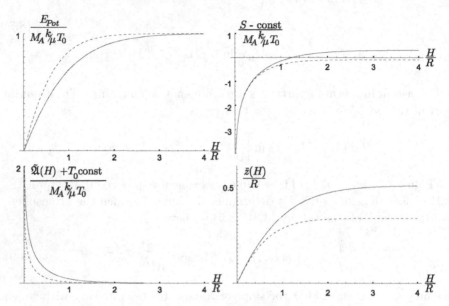

Fig. 8.1 Potential energy, entropy and available free energy of an atmosphere enclosed under a roof of height H.
Also: The mean height of the atmosphere.
Solid: $\beta = 2$, Dashed: $\beta = 3$.

For $H \to \infty$ the available free energy $\tilde{\mathfrak{A}}$ approaches a constant. We may say that the contractive tendency of the potential energy eventually catches up with the expansive tendency of the entropy. Both cancel each other. The competition is stalemated. As a result of the stalemate the atmosphere should stay with the planet. Indeed, its mean height is finite even for $H \to \infty$ and it is smaller for bigger values of $\beta = \frac{gR}{k/\mu\,T_0}$.

8.3 Spherical planet with a potential of the form $\varphi = -\frac{\gamma M}{r}$

However, that is not the whole truth. Indeed, we have made some assumptions in (8.3) which are justified only in the immediate neighbourhood of the surface. More generally we must assume that the gravitational potential φ equals $-\frac{\gamma M}{r}$, where M is the mass of the planet, $r = R + z$ is the distance from its center, and γ is the gravitational constant. Also the infinitesimal volume element $\mathrm{d}\mathbf{x}$ of the atmosphere must be written as $4\pi r^2 \mathrm{d}r$. The atmosphere now fills a spherical shell of thickness H and R is the radius of the planet.

With this we obtain for $\tilde{\mathfrak{A}}$ in (8.2)

$$\tilde{\mathfrak{A}} = 4\pi \int\limits_{R}^{R+H} \rho\,(r) \left\{ -\frac{\gamma M}{r} + \frac{k}{\mu} T_0 \ln \frac{\rho\,(r)}{\rho\,(T_R, p_R)} \right\} r^2 dr. \qquad (8.11)$$

We proceed as before and obtain with $g = \frac{\gamma M}{R^2}$

$$\rho(r; H) = \frac{M_A}{4\pi R^3} \frac{e^{\beta x}}{I_2(h; \beta)}, \quad \text{where} \quad \beta = \frac{gR}{k/\mu T_0}, \quad x = \frac{R}{r}, \quad h = \frac{H}{R}, \qquad (8.12)$$

and where we have defined the integral $I_2\,(h; \beta)$ which is one of the class of integrals

$$I_A(h; \beta) = \int\limits_{\frac{1}{1+h}}^{1} e^{\beta u} \frac{1}{u^{A+2}} du. \qquad (8.13)$$

We cannot represent such integrals in terms of elementary functions of h and β, but Mathematica$^{\circledR}$ knows their values and — upon instruction — is capable of plotting them. Therefore we may be satisfied with implicit results in terms of the integrals I_A. We obtain

$$\bar{r}(H) = R \frac{I_3(h; \beta)}{I_2(h; \beta)}$$

$$E_{\text{Pot}}(H) = M_A \frac{k}{\mu} T_0 \left(-\beta \frac{I_1(h; \beta)}{I_2(h; \beta)} \right)$$

$$\tilde{S}(H) = \frac{E_{\text{Pot}}(H)}{T_0} + M_A \frac{k}{\mu} \ln I_2(h; \beta) + \text{const.} \qquad (8.14)$$

$$\tilde{\mathfrak{A}}(H) = -M_A \frac{k}{\mu} T_0 \ln I_2(h; \beta) - T_0 \cdot \text{const.}$$

The plots are shown in Fig. 8.2. By (8.14)$_4$ and (8.12) we see that (8.9) is still true, so that the slope of $\tilde{\mathfrak{A}}$ determines the force on the roof of the constrained atmosphere. Much of the previous discussion about the energetic and the entropic parts of that force remains valid. In particular, for $H \to 0$, i.e. when the atmosphere is crammed to the surface of the planet, the expansive entropic force is infinite and far outweighs the contractive energetic force.

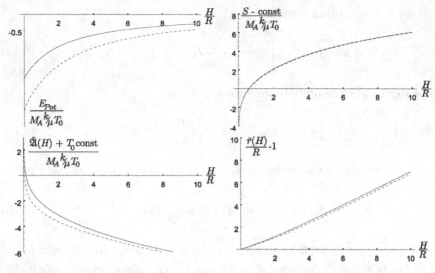

Fig. 8.2 The plots of Fig. 8.1 corrected by the consideration of the realistic potential $-\frac{\gamma M}{r}$ instead of the approximate potential gz. Solid: $\beta = 2$, Dashed: $\beta = 3$.

But there is now no stalemate for $H \to \infty$. Indeed, the available free energy does not approach a constant value now, rather if continues to decrease so that its minimum lies at $H \to \infty$. The entropy wins (!) and the atmosphere will be scattered throughout infinite space in equilibrium. This observation is also confirmed by the plot for $\bar{r}$: The mean height of the atmosphere tends to infinity when H goes to infinity. This means nothing else than that in equilibrium a planet has no atmosphere.

8.4 Escaping atmospheres

Actually we must not be too surprised, because among the molecules of the atmosphere there will always be some which exceed the escape velocity and they have a good chance of leaving the planet forever. Provided, of course, that they fly in the right direction and do not collide.

So why does the earth have an atmosphere? The answer must be that the earth's atmosphere is in a metastable state; it may take a long time for the whole atmosphere to escape. Inspection of the equations (8.14) shows that the relevant parameter is

$$\beta = \frac{gR}{k/\mu T_0} = \frac{\gamma \frac{M}{R}}{k/\mu T_0} \ . \tag{8.15}$$

In Figs. 8.1 and 8.2 we have shown the effect of this parameter upon $\tilde{\mathfrak{A}}$ by plotting *two* curves. The solid one for $\beta = 2$ and the dashed one for $\beta = 3$.[2] In the latter case $\tilde{\mathfrak{A}}(H)$ is flatter so that the expansive pressure of the atmosphere is smaller and presumably it will then take more time to strip the planet of its atmosphere.

Thus we see that

• large planetary masses, • small radii, • small temperatures

are conducive to a (meta)stable atmosphere, while the opposite is a recipe for a bare planet. This is why Mercury, a hot planet of little mass, has no atmosphere left, while Jupiter, a massive cold planet, has a thick and dense atmosphere.[3]

Actually the outer planets — Jupiter, Neptune, Uranus — are so massive, *because* they are cold. Let us consider: We see from (8.15) that the atomic or molecular mass μ also affects the size of β, so that β tends to be small for a small μ. Therefore the light gases in the atmosphere escape first. But a cold planet can even keep the light gases, and that is what happened to Jupiter. It has held on even to hydrogen and helium and since those elements were predominant when the planetary system was formed, the planet is massive.

Earth stands in the middle between Mercury and Jupiter. It is so warm that it has lost its free hydrogen and its helium, but it is just cold enough to hang on to oxygen and nitrogen — at least for a while.

A theory of equilibria — like the present one — gives no indication about the speed with which an eventual equilibrium establishes itself. Our experience with life on earth suggests that the relevant time scale for earth may be billions of years.

---

[2] For earth β has a value of approximately 800. For such large values the graphs of Figs. 8.1 and 8.2 are difficult to compare. And anyway we are after qualitative results here with our consideration of the model atmosphere.

[3] It is true that Jupiter also has a large radius, but in its case the large mass outweighs the large radius.

Entropy of mixing. Osmosis

Mixing of different constituents is a process associated with a considerable increase of entropy. In fact, entropy opposes de-mixing strongly and
complete separation absolutely. The entropic tendency for mixing may be
interfered with by the tendency of potential energy to decrease. The Pfeffer
tube furnishes an instructive example for such a competition between the
gravitational potential energy and entropy. At the same time the Pfeffer
tube may be considered as a model for the "mechanism" by which the sap
in the capillary ducts of a tree reaches the tree tops.

9.1 Mixing

Mixing of initially pure constituents – gases or liquids, or solids – is an insis-
tent and powerful process which is driven, primarily, by an increase of entropy.
To be sure, energy may also decrease by mixing, but even if energy increases,
the fact can never prevent mixing altogether, although it may interfere with
it. We proceed to investigate the process.

In order to fix the ideas we consider the situation shown in Fig. 9.1. Initially
we have ν pure constituents α – all at temperature T, pressure p and in the
individual volumes V_α – and they are separated by closed taps. The total
volume is $V = \sum_{\alpha=1}^{\nu} V_\alpha$. When the taps are opened, mixing starts – at constant
T and p – and we know that it may proceed until a homogeneous mixture
prevails in the total volume.[1] The mixing process is generally accompanied
by a change of total volume V, internal energy U, enthalpy $H = U + pV$ and
entropy S. We write these changes as

---

[1] This does not always happen, but let us take simple cases first.

$$V_{\text{Mix}} = m_\gamma v_\gamma(T, p_\beta) - \sum_{\alpha=1}^{\nu} m_\alpha v_\alpha(T, p) \quad \forall \ \gamma = 1, 2 ... \nu$$

$$U_{\text{Mix}} = \sum_{\alpha=1}^{\nu} m_\alpha \left(u_\alpha(T, p_\beta) - u_\alpha(T, p) \right)$$

(9.1)

$$H_{\text{Mix}} = U_{\text{Mix}} + pV_{\text{Mix}}$$

$$S_{\text{Mix}} = \sum_{\alpha=1}^{\nu} m_\alpha \left(s_\alpha(T, p_\beta) - s_\alpha(T, p) \right).$$

They are called the volume of mixing, internal energy of mixing, heat of mixing and entropy of mixing, respectively.

m_α are the masses of the constituents and p_β are the partial pressures of the constituents after mixing, such that $p = \sum_{\alpha=1}^{\nu} p_\beta$ holds. Note that the specific volumes v_α, the specific internal energies u_α and the specific entropies s_α after mixing may depend on all partial pressures $p_1, p_2, ... p_\nu$.

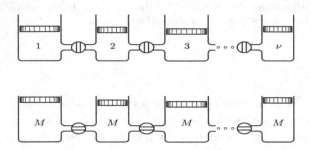

Fig. 9.1 Pure constituents at T, p before mixing (top).
Homogeneous mixture at T, p (bottom). [Note that the
volume may have changed in the mixing process.]

9.2 Mixing of ideal gases

However, by Dalton's law, this is not the case for mixtures of ideal gases. In such mixtures the constituent quantities v_α, u_α and s_α depend only on "their own pressure" p_α and, moreover, the dependence is the same one as if the constituent were pure. Therefore we have, cf. (4.9)

$$V_{\text{Mix}} = 0, \quad U_{\text{Mix}} = 0, \quad H_{\text{Mix}} = 0, \quad \text{but } S_{\text{Mix}} = \sum_{\alpha=1}^{\nu} N_\alpha k \ln \frac{V}{V_\alpha}.$$

(9.2)

All mixing quantities vanish except for the entropy of mixing and, obviously, the entropy of mixing is due to the volume expansion of all constituents.

9.3 Ideal mixture of liquids

In contrast to mixtures of gases, if we let liquids mix, we cannot calculate the mixing quantities, because we do not, in general, know the thermal and caloric equations of state, neither for the pure liquids nor for liquids which are constituents of a mixture.[2] Therefore we *assume* that there are some such mixtures — the so-called *ideal mixtures* — for which the mixing properties are given by (9.2), just as if the constituents were ideal gases.

Concerning V_{Mix} and H_{Mix} such an assumption is easily put to the test by observing the mixing process: If no change of volume occurs during the process nor any absorption or release of heat, then V_{Mix} and H_{Mix} both vanish. But what about the entropy of mixing? The expression $(9.2)_4$ clearly exhibits — in the logarithmic V-dependence — its origin in ideal gases, cf. (1.20). So, how could it possibly be relevant for solutions? And yet, we may *motivate* $(9.2)_4$ for arbitrary mixtures from the Boltzmann formula $S = k \ln W$. Let us consider:

We have discussed Boltzmann's expression (3.36) for entropy by saying — among other things — that this expression may be extrapolated away from monatomic gases. So let us now extrapolate that expression to distributions of atoms or molecules of liquids, so as to provide us with an interpretation of the entropy of mixing.

Before mixing, the atoms or molecules of constituent α at any one time are occupying the points $\mathbf{x} \in V_\alpha$.[3] Thus, if $N_{\mathbf{x}}^\alpha$ atoms occupy point $\mathbf{x}$, the entropy S_α^S at the start reads

$$S_\alpha^S = k \ln \frac{N^\alpha!}{\prod\limits_{\mathbf{x}}^{P_\alpha^S} N_{\mathbf{x}}^\alpha!} = k \left(N^\alpha \ln N^\alpha - \prod\limits_{\mathbf{x}}^{P_\alpha^S} N_{\mathbf{x}}^\alpha \ln N_{\mathbf{x}}^\alpha \right), \qquad (9.3)$$

where the second equation is approximate; it results from the application of the Stirling formula. Since the situation before mixing is a stationary equilibrium, we expect that all $N_{\mathbf{x}}^\alpha$'s are equal so as to make S_α^S maximal.

---

[2] The equations of state may be — and are — measured, and tables of data are published for technically important solutions or alloys. This is an interesting quantitative information for the engineer. We are here, however, more interested in formulae that can be processed analytically.

[3] As always we assume that the points are discrete and that they may be occupied by more than one atom or molecule. We denote the number of points in V_α by P_α^S. Also, later, after mixing the number of points occupiable by molecules of constituent α will be denoted by P_α^F.

Thus $N_x^\alpha = N^\alpha / P_\alpha^S$ holds and, if P_α^S, the number of points in V_α, is proportional to V_α, with X_α as a factor of proportionality, we have

$$S_\alpha^S = k \left(N^\alpha \ln N^\alpha - P_\alpha^S \frac{N^\alpha}{P_\alpha^S} \ln \frac{N^\alpha}{P_\alpha^S} \right)$$

$$S_\alpha^S = k \ N^\alpha \ln (X_\alpha V_\alpha) \qquad\qquad \forall\ \alpha = 1, 2 ... \nu. \qquad (9.4)$$

This holds for all constituents at the start, i.e. before mixing. The factor X may be different for different constituents. Therefore we denote it by X_α.

When the mixing is finished we have by the same argument

$$S_\alpha^F = k \ln \frac{N^\alpha!}{\prod\limits_{x}^{P_\alpha^F} N_x^\alpha!} \qquad\qquad (9.5)$$

and the only difference between (9.3) and (9.5) is the difference in P_α^S and P_α^F; the latter is equal to $X_\alpha V$, where V is the total volume, jointly occupied by all constituents after mixing. Therefore we obtain in a manner analogous to (9.4)

$$S_\alpha^F = k \ N^\alpha \ln (X_\alpha V) \qquad\qquad \forall\ \alpha = 1, 2 ... \nu. \qquad (9.6)$$

We thus have

$$S_\alpha^F - S_\alpha^S = k \ \ln \left(\frac{V}{V_\alpha} \right)^{N_\alpha}, \qquad\qquad (9.7)$$

and, by considering all constituents

$$S^F - S^S = \sum_{\alpha=1}^{\nu} N_\alpha k \ln \frac{V}{V_\alpha} . \qquad\qquad (9.8)$$

This is the *same equation* as (9.2) for the entropy of mixing, even though in the present argument we have made no explicit use of ideal gas properties. This observation lends some credence to the assumption that the entropy of mixing in the form (9.2) is universally true — at least for the special case of ideal mixtures: Not only for mixtures of ideal gases, but also for solutions of liquids and even for solid alloys.

Inspection of (9.7) and comparison with $S = k \ln W$ shows that $W = (V/V_\alpha)^{N_\alpha}$ is the ratio of the number of possibilities for N_α atoms to fill the volume V homogeneously and that same number in V_α. With N_α in the order of magnitude of 10^{20} and $V/V_\alpha = 2$ (say), the quotient is a terrifically large number. Therefore the random motion of the atoms or molecules will see to it that we nearly always see homogeneously mixed constituents and almost never spontaneous unmixing, at least not in ideal mixtures.

9.4 Pfeffer tube

An instructive example for the power of the entropy of mixing is provided by the Pfeffer tube which is an instrument often kept in high school physics laboratories for the demonstration of the phenomenon of *osmosis*. A long tube with one end sealed with a semipermeable membrane − permeable for water − is dipped, with that end, into a water-filled basin, cf. Fig. 9.2. Since water can pass the membrane, the water level will be even throughout the system as shown in Fig. 9.2_{left}.

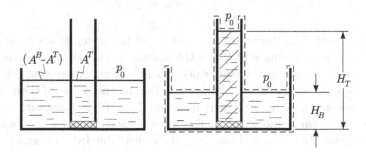

Fig. 9.2 Pfeffer tube.

Next we dissolve some substances with masses $m_\beta (\beta = 1, 2, ...\nu - 1)$ in the water inside the tube. The membrane is impermeable for those solutes. Therefore we expect the solutes to spread evenly throughout the little bit of water available to them and that should be the end of it, or so we might think. However, nature is cleverer than that. The solutes in their effort to increase their entropy try to spread over as big a volume as they can and therefore they "pull" water through the membrane; in this manner they obtain a larger volume to mix in and consequently the level of water rises in the tube. Or else, we may say that water "pushes" through the membrane[4] because in this manner it helps the solutes to increase their entropy.

Of course "pulling" and "pushing" are anthropomorphic euphemisms. What happens is the effect of thermal motion and blind chance that brings the system into *the* distribution of matter that has the largest number of realizations.

However, no matter whether the water is pulled or pushed into the tube, or whether blind chance is at work, the system will increase its potential energy when the water rises in the tube and that energetic expense will eventually put a stop on the entropic gain. Thus here we have another example for the competition of entropy and energy and we proceed to describe it analytically.

---

[4] The Greek word *osmos* means push.

We apply the 1st and 2nd laws to the control volume marked by the dashed line in Fig. 9.2$_{\text{right}}$. We assume that the temperature is homogeneous and constant and that the kinetic energy is negligible. The external working $\dot{A}_{\text{ext}}$ is equal to $-p_0 \frac{dV}{dt}$. Elimination of $\dot{Q}$ between the two laws provides an inequality for the rate of change of the available free energy $\mathfrak{A}$, viz.

$$\frac{d\left(U + E_{\text{Pot}} - TS + p_0 V\right)}{dt} \leqslant 0 . \tag{9.9}$$

The solution inside the tube is assumed ideal so that we have

$$V_{\text{Mix}} = 0, \quad U_{\text{Mix}} = 0, \quad S_{\text{Mix}} = \sum_{\alpha=1}^{\nu} N_\alpha k \ln \frac{V^T}{V_\alpha}, \tag{9.10}$$

where V_α ($\alpha = 1, 2, ...\nu$) denotes the volumes of the water in the tube and of the solutes before mixing and V^T is the volume of the solution in the tube. It is reasonable to assume all constituents before mixing as incompressible so that their specific volumes are constant and the specific values of their internal energies and entropies are only dependent on T, which is constant.

Thus, while the osmotic rearrangement of water between the basin and the tube occurs, the only variable part of the available free energy is

$$\tilde{\mathfrak{A}} = E_{\text{Pot}} - TS_{\text{Mix}} + p_0 V . \tag{9.11}$$

The rearrangement will stop, according to (9.9), when this quantity has reached its minimum. Note that the total volume is constant because of (9.10)$_1$.

We use indices as follows

T for the tube,
B for the basin,
$\beta = 1, 2, ...\nu - 1$ for the solutes,
ν for the solvent (water)

and write $\tilde{\mathfrak{A}}$ in the form

$$\tilde{\mathfrak{A}} = \frac{g}{2} \left(\rho^T A^T H_T^2 + \rho_\nu \left(A^B - A^T\right) H_B^2\right) -$$
$$- kT \left\{ N_\nu^T \ln \frac{A^T H_T}{V_\nu} + \sum_{\beta=1}^{\nu-1} N_\beta \ln \frac{A^T H_T}{V_\beta} \right\} + \tag{9.12}$$
$$+ p_0 (A^T H_T - V_\nu).$$

ρ_β and ρ_ν are the densities of the pure solutes and pure solvent respectively, while ρ^T is the density of the solution and ρ_ν^T the density of the solvent in the solution. Since we assume that the solution is ideal, so that V_{Mix} vanishes, a little calculation shows that we have

$$\rho_\nu^T = \rho_\nu \frac{1}{1 + \dfrac{\sum\limits_{\beta=1}^{\nu-1} V_\beta}{V_\nu}} \approx \rho_\nu$$

$$\rho^T = \rho_\nu \frac{1}{1 + \dfrac{\sum\limits_{\beta=1}^{\nu-1} V_\beta}{V_\nu}} + \sum_{\beta=1}^{\nu-1} \frac{m_\beta}{A^T H_T} \approx \rho_\nu + \sum_{\beta=1}^{\nu-1} \frac{m_\beta}{A^T H_T} \tag{9.13}$$

$$V_\nu = \frac{V^T}{1 + \dfrac{\sum\limits_{\beta=1}^{\nu-1} V_\beta}{V_\nu}} \approx A^T H_T$$

$$H_B = \frac{m_\nu}{\rho_\nu (A^B - A^T)} - \frac{A^T}{A^B - A^T} \frac{\rho_\nu^T}{\rho_\nu} H_T \approx \frac{1}{A^B - A^T}\left(\frac{m_\nu}{\rho_\nu} - A^T H_T\right).$$

The approximations in (9.13) refer to a dilute solution in which we have

$$\frac{\sum\limits_{\beta=1}^{\nu-1} V_\beta}{V_\nu} \ll 1 \qquad\qquad (\beta = 1, 2...\nu - 1).$$

We restrict the attention to this approximate case, because it exhibits all interesting phenomena which we wish to discuss.[5] The relevant part $\tilde{\mathfrak{A}}$ of the available free energy $\mathfrak{A}$ thus reads

$$\tilde{\mathfrak{A}} = \frac{g}{2}\left\{ \rho_\nu \frac{A^T A^B}{A^B - A^T} H_T^2 + \left(\sum_{\beta=1}^{\nu-1} m_\beta - 2m_\nu \frac{A^T}{A^B - A^T}\right) H_T + \frac{m_\nu^2}{\rho_\nu (A^B - A^T)} \right\} -$$

$$- T \sum_{\beta=1}^{\nu-1} N_\beta k \ln \frac{A^T H_T}{V_\beta}. \tag{9.14}$$

The graphs in Fig. 9.3 represent $\tilde{\mathfrak{A}}$ as well as E_{Pot} and S_{Mix}, all as functions of H_T. In order to prepare the graphs we have chosen parameters as follows

$$g = 9.81 \frac{m}{s^2}, \quad T = 298\ K, \quad m_\nu = 2\ \text{kg}, \quad A^B = 200\ \text{cm}^2, \quad A^T = 1\ \text{cm}^2.$$

As solute we have taken 1 gr of ordinary table salt $NaCl$. Since $NaCl$ dissociates into ions Na^+ and Cl^- in water we thus have *two* solutes with the same values N_β and V_β $(\beta = 1, 2)$, viz.

---

[5] Moreover, only dilute solutions have a chance of being ideal.

$$N_\beta = 1.02 \cdot 10^{22} \quad \text{and} \quad V_\beta = 0.46 \text{ cm}^3.$$

The available free energy has a minimum at $H_T = 9.3$ m (!) This marks the height to which the solution will rise in the tube.

We conclude that in this example nearly half of the water pushes its way from the basin into the tube. The entropy S_{Mix} increases in this manner, cf. Fig. 9.3, and that increase is exclusively due to the entropy of the salt in our approximation. The potential energy E_{Pot} increases parabolically and that increase is essentially due to a growth of the potential energy of the water. So, here is an interesting aspect of the competition of energy and entropy: The salt profits in the osmotic process by increasing its entropy, while the water has to pay for it, because its potential energy rises. However, we must not think of the constituents as benefactor and beneficiary; the *system* benefits by moving into the distribution with most realizations and the driving force is *chance*. Chance drives the thermal motion of the molecules to explore the system in all its microstates and *the* distribution of water and salt will be chosen that can be realized by most microstates.

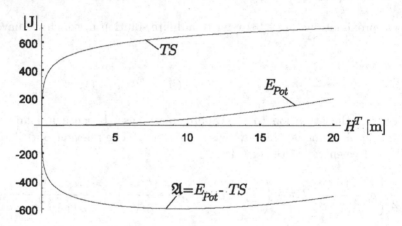

Fig. 9.3 E_{Pot}, TS_{Mix} and $\tilde{\mathfrak{A}}$ as functions of H_T[m].

A detail not visible in Fig. 9.3 is a minimum of E_{Pot} at

$$H_T^{\text{Min}} = \frac{1}{\rho_\nu A^B}\left(m_\nu - \frac{1}{2}\sum_{\beta=1}^{\nu-1} m_\beta \frac{A^B - A^T}{A^T}\right) = 0.095 \text{ m}. \tag{9.15}$$

Thus when the salt has gone into solution inside the tube the minimum of potential energy occurs when the water level in the tube is 5 mm below the 10 cm mark at which it stood when there was no salt yet.

An obvious detail of the graphs of Fig. 9.3 is that the entropic driving force is very strong when the osmotic push starts; this follows from the very steep increase of S_{Mix} for small H_T which entails a very steep decrease of $\tilde{\mathfrak{A}}$.

The hydrostatic pressure jump across the semipermeable membrane in equilibrium is equal to

$$P = \rho^T g H_T - \rho_\nu g H_B = 0.903 \text{ bar.}$$

This is the so-called osmotic pressure of the solution.

The rising of the solution in the Pfeffer tube is supposed to be a model for the supply of sap to the tree tops. The roots of the tree lie in the relatively pure ground water and their surface membranes are permeable for the water so that the water can enter the narrow ducts leading from the roots to the tree tops thus diluting the sap inside those ducts. It has been estimated that the osmotic effect can overcome a height difference of 100 m.

10

Phase transition

If energy and entropy compete in arranging the mass of a body, the energy need not be the potential energy of the gravitational field. It may be the potential energy of the intermolecular van der Waals forces, cf. Sect.4.4.

The non-convex character of the potential energy creates a qualitative difference between the effects of gravitational energy and of molecular interaction: The transition between the energy-dominated low temperature situation and the entropy-dominated high temperature case is no longer smooth for the particle interaction; rather it occurs abruptly at one temperature and one pressure in a phase transition.

10.1 Effect of molecular interaction

The atoms or molecules of a body offer each other numerous potential wells, in which they may rest comfortably, having a minimum of potential energy, and filling a small volume, because all particles are close together. Actually that situation occurs in all bodies at low temperature or subject to a high pressure, when they are in the solid or liquid phase.

When a liquid body is heated, the temperature rises and so does the mean kinetic energy of the particles. Eventually they will be fast enough to jump out of the potential wells and to explore the "outer space"; the body becomes vaporous. At a given pressure p this *phase transition* happens at a fixed temperature T_t and the energy required to lift all particles out of their potential wells is called the *heat of evaporation* $r(T_t)$, cf. (4.3).

The vapour requires much more space than the liquid, so that its entropy is bigger. And indeed, a phase transition may be seen as one aspect of the universal competition of energy tending toward a minimum, and of entropy tending toward a maximum. The evaporation of a liquid — or its opposite, the condensation of vapour — may be viewed as a sudden overpowering of one of these tendencies by the other one.

We proceed to investigate this interesting phenomenon from several points of view.

10.2 Molecular dynamics of the phase transitions solid ↔ liquid ↔ vapour

As a first demonstration of the phenomena of melting and evaporation – or of condensation and freezing – we have prepared an animation of a few interacting particles, as few as seven, which are enclosed in a fixed volume. Upon heating or cooling by contact with the wall, the particles will simulate a gas with free flying atoms at high temperatures, or a liquid in which the atoms cluster together at intermediate temperatures, or a hexagonal solid lattice with all next neighbours having the same distance; the latter case occurs at small temperatures. Figure 10.1 shows screen shots of the three phases. The full animation may be viewed on the accompanying CD in the program "solidification". The number of particles may be chosen, and the viewer has control over the heating and cooling stages.

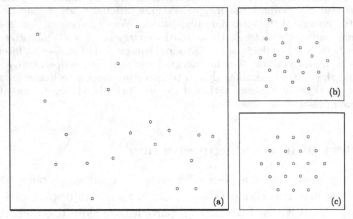

Fig. 10.1 Nineteen atoms simulating a gas (a), a liquid (b) and a
          solid (c). Screen shots from the program "solidification"
          of the accompanying CD.

10.3 Phase equilibrium

Next we consider a body enclosed in a cylinder with a movable piston. The temperature of the boundary is kept fixed at T_0 and the pressure on the piston is fixed at p_0, cf. Fig. 10.2. For simplicity we ignore the gravitational potential energy and, in particular, we consider the piston as weightless. Under those conditions the 1st and 2nd laws imply:

$$\mathfrak{A} = U - T_0 S + p_0 V \quad \rightarrow \quad \text{minimum in equilibrium.} \qquad (10.1)$$

The kinetic energy K is dropped for the reason given in Sect. 7.6, so that it remains to find the fields $T(\mathbf{x})$ and $v(\mathbf{x})$ that minimize

$$\mathfrak{A} = \int_m \{u(T(\mathbf{x}), v(\mathbf{x})) - T_0 s(T(\mathbf{x}), v(\mathbf{x})) + p_0 v(\mathbf{x})\}\, \mathrm{d}m.$$

Here we have represented $\mathfrak{A}$ as an integral over the specific values u, s and v. The integration is over the total mass. It is slightly easier to do that than deal with volume integrals as we did in Sect. 7.6. The simplification is due to the fact that the conservation of mass is taken into account implicitly rather than by a Lagrange multiplier. The results of both methods are identical, however.

With the local Gibbs equation $T\mathrm{d}s = \mathrm{d}u + p\mathrm{d}v$ we obtain

$$T(\mathbf{x}) = T_0 \quad \text{and} \quad p(\mathbf{x}) = p_0 \tag{10.2}$$

so that $T(\mathbf{x})$ and $p(\mathbf{x})$ are homogeneous in equilibrium.

This does not necessarily mean that the mass density $\rho(\mathbf{x})$ – or the specific values $u(\mathbf{x})$, $s(\mathbf{x})$ – are also homogeneous, although all of them are locally functions of T and p. However, in different points $\rho(T,p)$, $u(T,p)$ and $s(T,p)$ may be different functions of T and p. This happens when the body in the cylinder consists of different phases, say liquid and vapour, cf. Fig. 10.2, or solid and liquid. In a case like that we may ask what else – apart from T and p – is unchanged, if anything, when a mass-element passes from one phase to the other. The answer is: g, the specific free enthalpy. Indeed, the local Gibbs equation $T\mathrm{d}s = \mathrm{d}u + p\mathrm{d}v$ may be written as $\mathrm{d}g = -s\mathrm{d}T + v\mathrm{d}p$ and, if T and p do not change, nor does g.

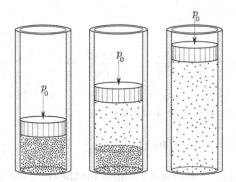

Fig. 10.2 A fluid under a given pressure p_0
and at a fixed temperature T_0.

It follows that $T(\mathbf{x})$, $p(\mathbf{x})$ *and* the specific free enthalpy $g(\mathbf{x})$ are all homogeneous in equilibrium, although neither $\rho(\mathbf{x})$, nor $u(\mathbf{x})$, or $s(\mathbf{x})$ are homogeneous.

For a phase equilibrium between liquid and vapour we thus have

$$g^L(T,p) = g^V(T,p), \tag{10.3}$$

which implies that, for a given T, phase equilibrium can only occur for one value of p. The function $p = p(T)$ implied by (10.3) is called the vapour pressure function.

In order to determine the vapour pressure function we need to know the functions $g^L(T,p)$ and $g^V(T,p)$ and this knowledge must be found experimentally, cf. Chap. 4. Therefore we do not usually have an analytic function $p(T)$. However, now we proceed to calculate such a function, albeit for a strongly simplified model.

10.4 Competition of energy - or enthalpy - and entropy

In order to exhibit the competition of energy and entropy in effecting a phase transition we assume that the liquid is incompressible and that the vapour is an ideal gas. Quantitatively that model is no good — particularly at or around the critical point — but heuristically the argument is neat. By (4.8) through (4.10) we have

$$h^L(T,p) = h(T_R,p_R) + c(T - T_R) + (p - p_R)v \tag{10.4}$$

$$h^V(T,p) = h(T_R,p_R) + c(T(p_R) - T_R) + r(T(p_R)) + (z+1)\frac{k}{\mu}(T - T(p_R))$$

for the enthalpies in $g = h - Ts$. And for the entropies we have

$$s^L(T,p) = s(T_R,p_R) + c\ln\frac{T}{T_R}$$

$$s^V(T,p) = s(T_R,p_R) + c\ln\frac{T(p_R)}{T_R} + \frac{r(T(p_R))}{T(p_R)} + (z+1)\frac{k}{\mu}\ln\frac{T}{T(p_R)} - \frac{k}{\mu}\ln\frac{p}{p_R}. \tag{10.5}$$

We recall that the reference values (T_R, p_R) are $(298\,\text{K}, 1\,\text{atm})$ and we choose $z = 3$, appropriate for a 3-atomic gas, and

$$T(p_R) = 373\ \text{K}, \quad r(T(p_R)) = 2257\frac{\text{kJ}}{\text{kg}}, \quad c = 4.18\frac{\text{kJ}}{\text{kgK}}, \tag{10.6}$$

which are the appropriate *measured* values for the temperature of evaporation of water at 1 atm and the corresponding heat of evaporation of water. c is the specific heat of liquid water. Insertion of these values implies for the enthalpy difference $(h^V - h^L)$ and entropy difference $(s^V - s^L)$, both as functions of $\vartheta = T/T(p_R)$ and $\pi = p/p_R$.

$$\frac{h^V - h^L}{874\frac{\text{kJ}}{\text{kg}}} = -(\vartheta - 1) + 2.58 - 1.14 \cdot 10^{-4}(\pi - 1),$$

$$\frac{T(s^L - s^V)}{874\frac{\text{kJ}}{\text{kg}}} = -\vartheta\ln\vartheta + 2.58\,\vartheta - 0.196\,\vartheta\ln\pi\ . \tag{10.7}$$

The interesting ranges of ϑ and π are

$$0.75 < \vartheta < 1.75 \quad \text{and} \quad 0.01 < \pi < 200,$$

because the upper and lower bounds define — approximately — the triple point and the critical point of water. [Of course, our model does not know those points.]

In those ranges we have $h^V > h^L$, so that the energy favours the liquid, and $s^V > s^L$, so that the entropy favours the vapour. In order to decide which phase prevails, we must know whether the energetic gain $(h^V - h^L)$ during the transition vapour $\rightarrow$ liquid is bigger or smaller than the entropic gain $T(s^V - s^L)$. Figure 10.3$_{\text{left}}$ allows us to recognise the situation.

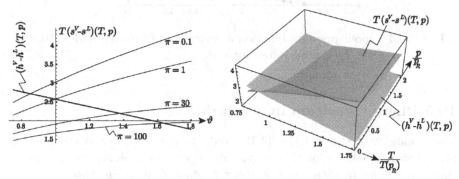

Fig. 10.3 Left: $(h^V - h^L)$ and $T(s^V - s^L)$ as functions of T and p. [Note that the π-dependence of $h^V - h^L$ is so weak that it does not show in the figure.]
Right: Ditto in a 3D-plot.

For T small, the gain of energy $(h^V - h^L)$ is bigger than the gain of the entropy term $T(s^V - s^L)$ for all relevant pressures p so that the liquid prevails. However, when T grows, the situation is reversed and vapour prevails; it does so earlier — i.e. for lower values of T — the smaller the pressure is, because a large pressure diminishes $(s^L - s^V)$, cf. (10.7) and Fig. 10.3.

Another aspect of this phenomenon can be read off from Fig. 10.3$_{\text{right}}$. The figure shows the surfaces $(h^V - h^L)$ and $T(s^V - s^L)$ according to (10.7) as functions of T and p and we confirm that low pressure and high temperature favour the vapour phase. The line of intersection of the two surfaces, projected onto the (p, T)-plane, defines the vapour pressure function $p = p(T)$. In the (p, T)-plane that function is a monotonically increasing convex graph. The analytical form of the graph is given by

$$\pi = \frac{1}{\vartheta^{5.1}} \exp\left\{18.26(1 - \frac{1}{\vartheta})\right\}. \tag{10.8}$$

Fig. 10.4 shows the graph of this function and also — as dots — some measured values of the vapour-pressure for water. We conclude that our model

– as expected – is quantitatively not good but qualitatively it gives reasonable results.

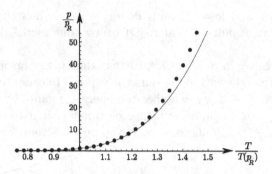

Fig. 10.4 Vapour-pressure function as calculated from the model (10.8). Dots: Measured values for water.

10.5 Phase transition in a van der Waals gas

The equations of state (4.11), (4.13) and (4.14) of van der Waals are the prototypical equations for the analytic description of a phase transition and they can provide us with a better understanding of the phenomenon.

We proceed to explain the situation by considering a van der Waals gas in a cylinder under a piston and with boundary values T_0 and p_0 for temperature and pressure. The available free energy appropriate to this case is $\mathfrak{A} = U + p_0 V - T_0 S$ and we write its energetic and entropic parts. These are

$$
U + p_0 V = \int_m \left[-a \left(\frac{1}{v(\mathbf{x})} - \frac{1}{v_R} \right) + \int_{T_R}^{T(\mathbf{x})} c_v(\beta) \mathrm{d}\beta + u_R + p_0 v(\mathbf{x}) \right] \mathrm{d}m
$$

(10.9)

$$
T_0 S = T_0 \int_m \left[\frac{k}{\mu} \ln \frac{v(\mathbf{x}) - b}{v_R - b} + \int_{T_R}^{T(\mathbf{x})} \frac{c_v(\beta)}{\beta} \mathrm{d}\beta + s_R \right] \mathrm{d}m \; .
$$

To begin with we imagine a case of uniform temperature $T(\mathbf{x}) = T_0$ and uniform specific volume $v(\mathbf{x})$ whose value v is such that it does *not* satisfy equilibrium of pressures, i.e. we have $p(v, T) \neq p_0$. Thus the case is a transient situation in which v changes in time while the fluid approaches the minimum of $\mathfrak{A}$.

Ignoring the constant and purely temperature-dependent terms in (10.9), we obtain for $\tilde{\mathfrak{A}}_E$ and $\tilde{\mathfrak{A}}_S$, the relevant – i.e. v-dependent – energetic and entropic parts of $\mathfrak{A}$

$$\tilde{\mathfrak{A}}_E = m\left[-a\frac{1}{v} + p_0 v\right] \quad \text{and} \quad \tilde{\mathfrak{A}}_S = -m\frac{k}{\mu}T\ln\left(\frac{v-b}{v_R}\right). \quad (10.10)$$

We plot these functions in Fig. 10.5 (a) and observe that the energy favours a small volume, the liquid, because the energetic minimum occurs for small values of v.

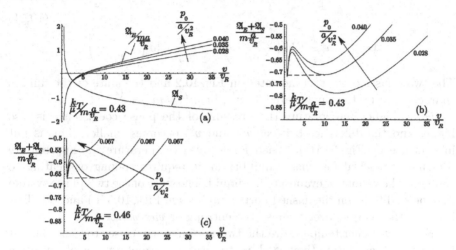

Fig. 10.5 (a) Energetic and entropic contributions to the available
free energy.
(b) The non-convex available free energy for three
different pressures p_0.
(c) The pressure of coexisting phases grows with increasing
temperature and the volume difference $v^V - v^L$ becomes
smaller.

Entropy, however, favours the vapour because its maximum lies at large volumes, cf. Fig. 10.5 (a). Thus energy and entropy favour different distributions of mass just as they did in Chap. 8 and 9, where we considered the planetary atmosphere and the Pfeffer tube. However, there is a novelty in the present case. Indeed, in those previous cases the available free energy was a convex function, while here the sum of $\tilde{\mathfrak{A}}_E$ and $\tilde{\mathfrak{A}}_S$ is non-convex with two minima, at least in the interesting range of T and p_0, cf. Fig. 10.5 (b). Let us consider this new situation.

It is reasonable to assume that the fluid in the cylinder arranges its density so as to allow the available free energy to occupy the *deeper* minimum. For a small pressure we may therefore conclude that the fluid is in the vapour phase, where the volume is large. In contrast to that a high pressure favours the liquid phase with a small volume, cf. Fig. 10.5 (b).

At some intermediate pressure the minima are equally deep and that is where the liquid and vapour volumes are equally favourable; at *one* temperature and *one* pressure. Therefore both phases will form, and the field $v(\mathbf{x})$

can no longer be homogeneous. Rather it assumes the two values v^L and v^V given by the abscissae of the minima. By (10.9) the relevant energetic and entropic parts of the available free energy thus read

$$\tilde{\mathfrak{A}}_E = m^L \left[-a\frac{1}{v^L} + p_0 v^L \right] + m^V \left[-a\frac{1}{v^V} + p_0 v^V \right],$$

(10.11)

$$\tilde{\mathfrak{A}}_S = -m^L \frac{k}{\mu} T \ln \left(\frac{v^L - b}{v_R} \right) - m^V \frac{k}{\mu} T \ln \left(\frac{v^V - b}{v_R} \right).$$

The two phases coexist as indicated in Fig. 10.2 and the state of the fluid is characterized by the single variable m^V, or $m^L = m - m^V$.

For a higher temperature the pressure of the phase coexistence is also higher and the difference between v^L and v^V becomes smaller. This is put in evidence by Fig. 10.5 (c). When for a given temperature the appropriate pressure is reached for phase equilibrium, it requires cooling or heating to condense the vapour or evaporate the liquid. The state of the two-phase system may be said to lie on the dashed horizontal lines in Figs. 10.5 (b) and (c). This is where the energy "overwhelms" the entropy or *vice versa*.

For a yet higher temperature the two minima grow closer together and at the *critical temperature* they combine to form a horizontal point of inflection in the $\mathfrak{A}(v)$-curve. The engineer knows all this, of course. Indeed, he is likely to be well-acquainted with the (p, v)-diagram, or the (h, s)-diagram and their wet-steam-regions of coexisting boiling liquid and saturated vapour. Those regions become narrower as the temperature grows and vanish altogether at the critical point. However, the engineer does not usually associate these phenomena with a competition between energy and entropy.

Before the days of van der Waals it was not generally accepted that all fluids have a critical point. To be sure, the phenomenon was known, because there were some fluids, like CO_2, where the critical point occurs virtually at room temperature; but it required the analysis of van der Waals to convince the physicists and engineers that the critical point is a universal phenomenon.

The very fact that vapour prevails at high temperature irrespective of the value of the pressure, although there are potential wells available to the atoms, is a case of *entropic stabilization of a high-temperature-phase*: The entropic gain of the vapour outweighs the energetic loss of the liquid. We shall come back to this phenomenon in more detail when we discuss martensitic transformations in solids.

10.6 The wet steam region; an exercise in molecular dynamics

The accompanying CD in the program "*pv*-diagram" has a simulation of an isothermal evaporation under a decreasing pressure. Along with the animation of the molecular motion the isotherms are plotted. The program takes a certain amount of time. But, if the operator is patient enough to wait for three or four isotherms to become complete, he will be rewarded with isotherms in the (p, v)-diagram which are complete in all major qualitative features: Steep isotherms in the liquid phase, horizontal branches in the wet steam region which increases in width for lower temperatures, and flat isotherms in the range of supersaturated steam. Figure 10.6 shows a screenshot after a five hour calculation in which we have indicated – by the solid line – the wet steam region.

We suggest that the reader play with the molecular dynamics of the program "pv-diagram" on the accompanying CD by changing the speed of the pressure drop or by changing the number of atoms of the gas. There are interesting details to be demonstrated about superheating and the visual formation of the coexisting liquid and vapour phases.

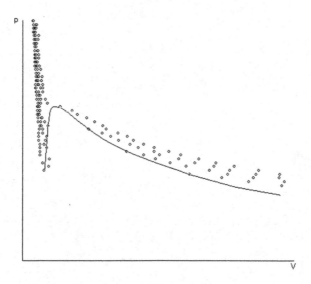

Fig. 10.6 Isotherms in the (p, v)-diagram and the wet steam region. Screenshot of an animation of the program "pv-diagram" on the accompanying CD.

Phase diagrams

Since ancient times smelting of ore has been a primary task of the foundry-men, and the modern chemical engineer faces a thermodynamically similar task when he refines oil and natural gas. Nature offers her resources mostly as mixtures and man mostly wants them in the pure form, or at least in enriched form. It has long since been learned that the recipe for enrichment is smelting or destillation, but the thermodynamic knowledge behind the tricks governing these processes are more recent. They consist of outwitting the entropy of mixing, and sometimes the heat of mixing can promote that purpose.

11.1 Available free energy in a mixture

We consider the same situation as often before: A fluid in a cylinder with boundary temperature T_0 closed by a piston which exerts the pressure p_0. Thus once again the available free energy is given as $\mathfrak{A} = U + p_0V + K - T_0S$ from which we drop the kinetic energy as irrelevant to the problem. The new aspect in this chapter is that now the cylinder shall contain a mixture of two constituents $\alpha = 1$ and $\alpha = 2$ with mol-numbers $\tilde{N}_\alpha$.

Chemical engineers and chemists — at least the anorganic types — have the "mol" as unit and, when they speak about specific values of energy (say) or entropy, those are molar values with the dimension J/mol or J/mol K. We shall denote those molar values by $\tilde{u}$ and $\tilde{s}$, respectively, so as to distinguish them from the mass-specific values u and s. The only other thing we have to know about the molar world is that the Avogadro number A now has the dimension $\frac{1}{\text{mol}}$ so that the mol-number $\tilde{N}_\alpha$ and the particle number N_α are related by

$$\tilde{N}_\alpha = N_\alpha/A$$

Thus $\tilde{N}_\alpha$ has the dimension mol. Also $\tilde{R} = Ak = 8.314$ J/mol K is the molar Boltzmann constant, often referred to as the ideal gas constant.

We may write the available free energy in the form

$$\mathfrak{A} = \int_{\tilde{N}} (\tilde{u}(\mathbf{x}) - T_0 \tilde{s}(\mathbf{x}) + p_0 \tilde{v}(\mathbf{x})) \mathrm{d}\tilde{N}, \tag{11.1}$$

where the integration is over the total mol number of the mixture and the integrand contains the molar values $\tilde{u}$, $\tilde{s}$ and $\tilde{v}$. As before in Chaps. 7 and 10 we assume *local equilibrium*. On those previous occasions we have expressed that assumption in the form $\mathrm{d}\tilde{s} = \frac{1}{T}(\mathrm{d}\tilde{u} + p\mathrm{d}\tilde{v})$ but in a mixture a new element needs to be taken care off. Indeed, in a mixture $\tilde{u}(\mathbf{x})$ and $\tilde{v}(\mathbf{x})$ are not the only variables characterizing the molar element $\mathrm{d}\tilde{N}$; there is also $X_1(\mathbf{x})$, the mol fraction of constituent 1, the fraction of mols of constituent 1 at the point $\mathbf{x}$. Therefore local equilibrium requires $\mathrm{d}\tilde{s} = \frac{1}{T}(\mathrm{d}\tilde{u} + p\mathrm{d}\tilde{v} - \tilde{\mu}\mathrm{d}X_1)$ or, more explicitly

$$\left(\frac{\partial \tilde{s}}{\partial T}\right)_{\tilde{v}, X_1} = \frac{1}{T}\left(\frac{\partial \tilde{u}}{\partial T}\right)_{\tilde{v}, X_1},$$

$$\left(\frac{\partial \tilde{s}}{\partial \tilde{v}}\right)_{T, X_1} = \frac{1}{T}\left[\left(\frac{\partial \tilde{u}}{\partial \tilde{v}}\right)_{T, X_1} + p\right], \tag{11.2}$$

$$\left(\frac{\partial \tilde{s}}{\partial X_1}\right)_{T, \tilde{v}} = \frac{1}{T}\left[\left(\frac{\partial \tilde{u}}{\partial X_1}\right)_{T, \tilde{v}} - \tilde{\mu}\right].$$

We know nothing about $\tilde{\mu}$ at this state, except that it is a function of T, $\tilde{v}$, X_1, or of T, p, X_1, because $\tilde{s}$ and $\tilde{u}$ are such functions.

We exploit the requirement that $\mathfrak{A}$ tends to a minimum as equilibrium is approached and consider the constraint

$$\tilde{N}_1 = \int_{\tilde{N}} X_1 \mathrm{d}\tilde{N} \tag{11.3}$$

by using a Lagrange multiplier λ_1. Thus we obtain

$$T(\mathbf{x}) = T_0, \qquad p(\mathbf{x}) = p_0, \qquad \tilde{\mu}(\mathbf{x}) = \lambda_1. \tag{11.4}$$

Thus T, p and $\tilde{\mu}$ are all homogeneous fields in equilibrium. Since $\tilde{\mu}$ is a function of T, p and X_1, (11.4) may seem to indicate that X_1 is also a homogeneous field in equilibrium.

This is not so, however, or at least it is not so in general. Indeed, $\tilde{u}$ and $\tilde{s}$, hence $\tilde{\mu}$ may be different functions of T, p and X_1 in different parts of the mixture. This infact happens when part of the mixture exists in different phases − liquid and vapour (say) − in different parts of the body. Within those phases all fields are homogeneous. Also T, p and and $\tilde{\mu}$ are homogeneous throughout the body but $\tilde{u}$, $\tilde{s}$ and X_1 are not. We may then ask for the fraction $z = \frac{\tilde{N}^V}{\tilde{N}}$ of mols in the vapour phase and for the mol fractions X_1^V and X_1^L of constituent 1 in the two phases in equilibrium.

In order to answer that question we assume that p and T are already homogeneous − with $T(\mathbf{x}) = T_0$ and $p(\mathbf{x}) = p_0$ − while constituent 1 still adjusts itself in the two phases so as to make $\tilde{\mu}$ homogeneous. Therefore, during that adjustment process the specific availability is equal to the free enthalpy $\tilde{g} = \tilde{u} - T\tilde{s} + p\tilde{v}$ and the availability itself may be written in the form

$$\mathfrak{A} = \tilde{N}z\tilde{g}^V(T, p, X_1^V) + \tilde{N}(1-z)\tilde{g}^L(T, p, X_1^L). \tag{11.5}$$

Equilibrium prevails for *the* values of z, X_1^V and X_1^L for which $\mathfrak{A}$ is minimal under the constraint

$$X_1 = X_1^V z + X_1^L(1-z). \tag{11.6}$$

We take care of the constraint by using a Lagrange multiplier λ and obtain

$$\lambda = \frac{\partial \tilde{g}^V}{\partial X_1^V} = \frac{\partial \tilde{g}^L}{\partial X_1^L} = \frac{\tilde{g}^V - \tilde{g}^L}{X_1^V - X_1^L}. \tag{11.7}$$

Therefore g^V and g^L must have a common tangent. In other words: The graph $min[g^V(X_1), g^L(X_1)]$ must be non-convex. The equilibrium values X_{1E}^V and X_{1E}^L are the abscissae of the points where the tangent touches g^V and g^L respectively, cf. Fig. 11.1. The vapour fraction z results from (11.6) as

$$z_E = \frac{X_1 - X_{1E}^L}{X_{1E}^V - X_{1E}^L} \tag{11.8}$$

and the availability of the two-phase body reads

$$\frac{\mathfrak{A}_E}{N} = g^L(T, p, X_{1E}^L) + \frac{X_1 - X_{1E}^L}{X_{1E}^V - X_{1E}^L}\left[g^V(T, p, X_{1E}^V) - g^L(T, p, X_{1E}^L)\right]. \tag{11.9}$$

This implies that the equilibrium state lies on the common tangent of the graphs $g^V(X_1^V; T, p)$ and $g^L(X_1^L; T, p)$. The reason is clear: That tangent has a lower free enthalpy than either one of the homogeneous phases in the range $X_1^V \le X_1 \le X_1^L$.

The construction of the common tangent is called *convexification* of the graph $min\,[g^V, g^L]$ and the equilibrium state is said to lie on the *convex envelope*.

11.2 Convexification of $g(T, p, X_1)$ with $\tilde{h}^h_{\text{Mix}}$ only

By Sect. 9.1 the molar free enthalpy $\tilde{g}^h(T, p, X_1^h)$ with $h = (V, L)$ is a sum of the free enthalpies $\tilde{g}_\alpha^h(T, p)$ of the pure constituents $\alpha = 1$ or 2 plus possible mixing terms. We may write

$$\tilde{g}^h(T, p, X_1^h) = \sum_{\alpha=1}^{2} X_\alpha^h \tilde{g}_\alpha^h(T, p) + \tilde{h}^h_{\text{Mix}} - T\tilde{s}^h_{\text{Mix}}. \tag{11.10}$$

The presence of $\tilde{h}_{\text{Mix}}$ and $\tilde{s}_{\text{Mix}}$ in (11.10) foreshadows another energetic-entropic competition in mixing which we proceed to clarify.

In order to be specific we take the free enthalpies $g_\alpha^h(T,p)$ of the pure constituents to be those of ideal gases for $h = V$ and those of incompressible liquids for $h = L$. Thus by (10.4) and (10.5) we have

$$\tilde{g}_\alpha^V(T,p) = (z_\alpha + 1)\tilde{R}[T - T_\alpha(p_R) - T \ln \tfrac{T}{T_\alpha(p_R)}] +$$

$$+ \tilde{R}T \ln \tfrac{p}{p_R} + \tilde{c}_\alpha \left[T_\alpha(p_R) - T_R - T \ln \tfrac{T_\alpha(p_R)}{T_R} \right] +$$

$$+ \tilde{r}_\alpha(T(p_R)) \left(1 - \tfrac{T}{T_\alpha(p_R)} \right) + \left(\tilde{h}_R^\alpha - T\tilde{s}_R^\alpha \right),$$

$$\tilde{g}_\alpha^L(T,p) = \tilde{c}_\alpha \left[T - T_R - T \ln \tfrac{T}{T_R} \right] + (p - p_R)\tilde{v}_\alpha + \left(\tilde{h}_R^\alpha - T\tilde{s}_R^\alpha \right). \qquad (11.11)$$

where $\tilde{h}_R^\alpha$, $\tilde{s}_R^\alpha$ stand for $\tilde{h}^\alpha(T_R, p_R)$, $\tilde{s}^\alpha(T_R, p_R)$. The term with the heat of evaporation $\tilde{r}$ dominates the T-dependence and therefore we conclude that at high temperatures $g_\alpha^V < g_\alpha^L$ holds for all X_1 while for low temperature we have $g_\alpha^L < g_\alpha^V$.

While there are mixtures in which the heat of mixing vanishes or may be neglected, the entropy of mixing is never negligible, not even in a mixture of ideal gases. And yet – for the sake of the argument – we now proceed to consider only $\tilde{h}_{\text{Mix}}^h$. In other words: we ignore $\tilde{s}_{\text{Mix}}^h$ altogether.

In the next section we draw $\tilde{s}_{\text{Mix}}^h$ into consideration. The reason for this procedure is that we wish to be able to appreciate the effects of energy and entropy upon the shape of phase diagrams separately.

The heat of mixing occurs when unequal neighbouring particles have a different interaction energy than equal ones. Thus, when unequal particles attract each other more strongly by the intermolecular forces than equal ones do, the potential well of the unequal pair is deep and the particles falling into the well accelerate. The temperature is raised or – when the temperature is kept unchanged – heat is released. This means that the heat of mixing is negative. In that case mixing is energetically encouraged. We are more interested in the opposite case: When the binding energy between two unlike particles is less than between like particles, there is an energetic penalty for mixing and heat must be added to achieve it.

Purely probabilistically the expected number of unlike pairs is equal to $N_1 \cdot \frac{N_2}{N}$, where N_α ($\alpha = 1, 2$) are the numbers of particles of the two mixing constituents and N stands for $N_1 + N_2$. Therefore it seems reasonable to consider the molar heat of mixing in each phase as proportional to $X_1(1 - X_1)$ with a factor of proportionality e. Thus we set

$$\tilde{h}_{\text{Mix}}^h = e^h X_1^h (1 - X_1^h), \qquad\qquad h = (V, L) \qquad (11.12)$$

In the sequel we take $e^V < 0$ and $e^L > 0$ and thus obtain convex and concave parabolae respectivley for $\tilde{g}^V(T, p, X_1^V)$ and $\tilde{g}^L(T, p, X_1^L)$, cf. Fig. 11.1. [Recall that, for the time being, $\tilde{s}^h_{\text{Mix}}$ equals zero.]

Once e^V and e^L have been chosen so as to suit our arguments, the T-dependence of the parabolae is dictated by the T-dependence of the end points $\tilde{g}^h_\alpha(T, p)$, cf. (11.11). [We shall keep p fixed once and for all.] Thus at a high temperature $\tilde{g}^V(T, p, X_1^V)$ is lower than $\tilde{g}^L(T, p, X_1^L)$ and the opposite is true for high temperature. Therefore the hot mixture will be in the vapour phase irrespective of the value of X_1 and the cold mixture will be liquid. Interesting are the intermediate temperatures where the curves intersect.

At intermediate temperatures convexification is possible as described in Sect. 11.1 and the mixture falls apart into two phases – a *pure* liquid and a vapour *mixture* – so that its state lies on the common tangent[1],where it achieves a lower free enthalpy than on either curve for the homogeneous mixture, cf. Fig.11.1$_{\text{left}}$. For lower temperatures there are two common tangents – or subtangents – which grow into a single one as the connecting line of the end points of $\tilde{g}^L$ touches the graph of $\tilde{g}^V$. From then on, always with decreasing temperature, the idea of convexification assumes a new meaning altogether, because the convex envelope connects the two end points of $\tilde{g}^L$ and this means that the mixture falls apart into two pure liquids.

It is common and convenient to summarize these phenomena in a (T, X_1)-*phase diagram* by projecting the subtangents onto the appropriate horizontal line $T = const$. The endpoints of all projections are then connected and what appears is a (T, X_1)-phase diagram with four regions, denoted by I through IV in Fig. 11.1$_{\text{left}}$. We have

    I.   homogeneous vapour mixture
    II. pure liquid 2 coexisting with a vapour mixture
    III. pure liquid 1 coexisting with a vapour mixture
    IV. pure liquids 1 and 2 coexisting

The region IV is called the *miscibility gap* for obvious reasons. The horizontal line on top of the miscibility gap is called the eutectic line with the eutectic point E. The eutectic temperature is the lowest temperature where evaporation occurs.

Arguments analogous to these for a liquid-vapour transition can be made for a solid-liquid transition. In that case T_E is the lowest melting temperature and that is where the name eutectic comes from. In ancient Greek this word means "easy melting".

---

[1] To be sure the tips in the end points of $\tilde{g}^L$ have many tangents. The important one is the one which touches $\tilde{g}^V$. Mathematicians working in convex analysis – if we understand them right – speak of a subtangent.

11.3 Phase diagram with $\tilde{h}^L_{\mathrm{Mix}}$ and $\tilde{s}^{L,V}_{\mathrm{Mix}}$

The foregoing consideration has shown us the energetic effect of the heat of mixing on a mixture and for the sake of a simple argument we have ignored the entropy of mixing. The entropy of mixing was previously calculated in (9.8). It is expressed in terms of the volume V of the mixture and the volumes V_α of the pure constituents, all under temperature T and pressure p. Under the assumption that the individual molecules of all constituents require the same volume both in the pure fluids and in the mixture we may write[2]

$$\frac{V}{V_\alpha} = \frac{N}{N_\alpha} = \frac{1}{X_\alpha} \quad \text{and therefore} \quad \tilde{s}_{\mathrm{Mix}} = -\sum_{\alpha=1}^{\nu} X_\alpha \tilde{R} \ln X_\alpha . \qquad (11.13)$$

In the present case we have $\nu = 2$ and the molar entropy of mixing assumes the form

$$\tilde{s}_{\mathrm{Mix}} = -\tilde{R}(X_1 \ln X_1 + (1 - X_1) \ln(1 - X_1)); \qquad (11.14)$$

it is universal and for $X_1 = 0, 1$ it is zero, of course, but its slope is infinite.[3]

When we consider $h^V_{\mathrm{Mix}} = 0$, h^L_{Mix} given by (11.12) and $\tilde{s}_{\mathrm{Mix}}$ for both phases as given by (11.14) and introduce those quantities into (11.10) we obtain the free enthalpies $\tilde{g}^V$ and $\tilde{g}^L$ as shown in the central column of Fig. 11.1. Inspection shows that $\tilde{g}^L$ is non-convex − due to the effect of h_{Mix}. But near $X_1 = 0, 1$ the graph is convex because of the steep slopes of $\tilde{s}_{\mathrm{Mix}}$ at those positions.

The interpretation of these graphs is much like the one of the graphs in Fig. 11.1$_{\mathrm{left}}$. But there are subtle differences, because the convexifying tangents touch the convex parts of the appropriate curves in points bounded away from $X_1 = 0, 1$. Therefore the (T, X_1)-phase diagram − still obtained by projection of the tangents − exhibits six regions, viz.

    I.   homogeneous vapour mixture,
    II. 2-rich liquid mixture called the homogeneous solution α,
    III. 1-rich liquid mixture called the homogeneous solution β,
    IV. two phases composed of vapour and solution α
    V. two phases composed of vapour and solution β
    VI. two liquid phases composed of the solutions α and β.

The latter is again called a miscibility gap.

Comparing the left and central columns of Fig. 11.1 we see what the addition of the entropy of mixing has done to the mixture: The entropy of mixing strongly oppposes unmixing and therefore it does not allow pure phases consisting of one constituent. To be sure the solutions α and β are 2-rich or 1-rich respectively but they are not pure.

---

[2] The assumption is true for ideal gases, but non generally for liquids and certainly not, if V_{Mix} is unequal to zero.

[3] The steep slopes cannot be detected in the graphs of Fig. 11.1, because they occur too closely to the lines $X^h_1 = 0, 1$.

The miscibility gap becomes wider when the temperature is lowered, because the bow-like entropy of mixing becomes flatter − because of its factor T in g^L, cf. (11.10) − so that the heat of mixing can push a bigger and wider concave bulge into that curve.

11.4 Phase diagram with $\tilde{s}_{\text{Mix}}^{L,V}$ only

If there is no heat of mixing, the phase diagram is quite different. Indeed all curves $\tilde{g}^V$ and $\tilde{g}^L$ are then convex, because their curvature is dictated by the entropy of mixing. We have drawn some such curves in the right column of Fig. 11.1. To be sure the graph $min[\tilde{g}^V, \tilde{g}^L]$ is non-convex and that permits convexification at intermediate temperatures. The phase diagram results from projection of the common tangents as before. If their end points are connected we obtain the two curves shown in the lower right corner of Fig. 11.1. Those curves are called the dew line and evaporation line respectively as indicated. The phase diagram has only three regions, viz.

I.   homogeneous vapour mixture
II. two-phases composed of vapour mixture and liquid mixture
III. homogeneous liquid solution.

The phase diagram is an important tool for the chemical engineer, because it helps him to design a destillation process for the separation of the constituents of a mixture or, better: For the enrichment of a constituent in a mixture. Let us consider:

We start at point I in the phase diagram of Fig. 11.2 with a homogeneous liquid mixture and increase the temperature T. Once the evaporation line is reached, a vapour bubble apears and that bubble has the mol-fraction X_1^{II} which is considerably smaller than X_1^I so that constituent 2 is enriched inside the bubble. Of course the liquid becomes richer in constituent 1 by releasing the bubble. Thus the temperature may rise a little before the next bubble appears which is not quite so rich in constituent 2 as the first one. If we continue to increase the temperature, the state of the liquid will move upwards on the evaporation line, while the state of the vapour − with all bubbles combined − moves up along the dew line. Eventually, when all is evaporated, the vapour has regained the mol-fraction X_1^I and − upon a further increase of temperature − moves up vertically in the phase diagram. The clever chemical engineer stops the process before complete evaporation has occured and thus comes away with a 2-rich vapour and a 1-rich liquid. Both may serve his purposes. If he needs a higher degree of enrichment, he has to repeat the process starting with the already enriched mixtures.

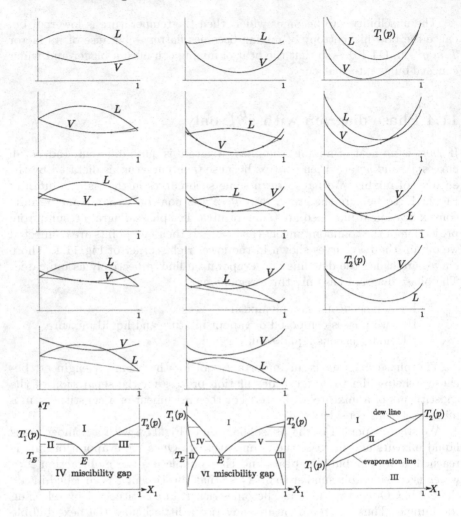

Fig. 11.1 Free enthalpies of vapour and liquid. Phase diagrams

Left:	Heats of mixing $h^V_{\text{Mix}} < 0$, $h^L_{\text{Mix}} > 0$. Entropies of mixing equal to zero
Center:	Heats of mixing $h^V_{\text{Mix}} = 0$, $h^L_{\text{Mix}} > 0$. Entropies of mixing cf. (11.14)
Right:	No heats of mixing. Entropies of mixing cf. (11.14). Temperature grows from *bottom to top*.

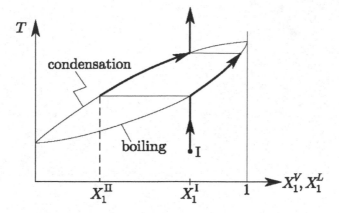

Fig. 11.2 An evaporation process in a mixture

Note that the engineer would be better served if there were neither entropy of mixing nor heat of mixing. In that case the liquid mixture upon evaporation would be separated into the pure vapour 2 and the pure liquid 1 for all temperatures between boiling temperatures $T_2(p)$ and $T_1(p)$. The entropy of mixing makes the life of the engineer difficult. But sometimes the heat of mixing may come to the rescue.

Indeed we have seen that the liquid phase "unmixes" to a considerable extent under the influence of a heat of mixing when its value is big enough. The solutions α and β separate and, since one is invariably lighter than the other one, it floats and may be scooped off — like the fat from the milk.

Chemical reactions

In chemical reactions the additive constants in the internal energies and entropies of the constituents play a significant role. To be sure, it is not the constants themselves that matter but certain combinations, which we may call the heat of reaction and the entropy of reaction. These values can be determined experimentally and have been tabulated.

The balance of the heat of reaction and the entropy of reaction goes a long way in determining whether a reaction can proceed and in which direction it proceeds. We give some examples: Dissociation of oxygen, ammonia synthesis and photosynthesis of glucose. The latter appears to be impossible, if we did not know better.

12.1 Competition of energy and entropy constants

The energy competing with entropy need not be the gravitational potential energy nor the potential energy of the interatomic van der Waals forces. It can be the binding energy of chemical bonds. When molecules of some constituents combine and those of other constituents split up, it is not only the masses of the constituents that appear and disappear; it is also the energies and the entropies and that includes the additive constants involved in the energies and entropies. Thus, if the energy constant of a reaction product is large and the sum of the energy constants of the reactants is small, the reaction has some difficulty to proceed. On the other hand the reaction is facilitated, if the reactants have a smaller entropy constant than the reaction products.

Thus whether and under which conditions a reaction occurs, is again a matter of competition of energy and entropy and the outcome is strongly affected by the values of the additive constants. We proceed to discuss this quantitatively.

12.2 A modicum of stoichiometry

We consider homogeneous reactions in a "well-stirred tank" of volume V, in which we maintain a constant and homogeneous temperature T by heating, or cooling, cf. Fig.12.1. The piston guarantees a constant pressure p which we assume to be homogeneous, because the effect of body forces on the content of the tank is neglected. Inside the tank we have ν constituents $\alpha = 1, 2...\nu$ with the mol-numbers $\tilde{N}_\alpha$ which may change on account of a *single* chemical reaction. In a formal manner we write the reaction in the form

$$\sum_{\alpha=1}^{a} \gamma_\alpha \alpha \rightarrow \sum_{\alpha=a+1}^{\nu} \gamma_\alpha \alpha \,,$$

where $\alpha = 1, 2, ...a$ denotes the reactants and $\alpha = a + 1, ...\nu$ denotes the reaction products. The γ's are the stoichiometric coefficients; we choose them as negative numbers for the reactants $\alpha = 1, ...a$ and as positive numbers for the products $\alpha = a + 1, ...\nu$. Also we normalize the stoichiometric coefficients by setting $\gamma_1 = -1$. If λ is the reaction rate density, the equations of balance of mol-numbers $\tilde{N}_\alpha$ read

$$\frac{d\tilde{N}_\alpha}{dt} = \gamma_\alpha \lambda V, \quad \text{hence} \quad \tilde{N}_\alpha(t) = \tilde{N}_\alpha(0) + \gamma_\alpha \Re \,, \qquad (12.1)$$

where $\Re = \int_0^t \lambda V \, dt$ is called the *extent of reaction*. $\Re$ like $\tilde{N}_\alpha$ has the dimension "mol".

From (12.1) we obtain

$$\frac{\tilde{N}_\alpha}{\gamma_\alpha} - \frac{\tilde{N}_\beta}{\gamma_\beta} = \frac{\tilde{N}_\alpha(0)}{\gamma_\alpha} - \frac{\tilde{N}_\beta(0)}{\gamma_\beta} \,. \qquad (12.2)$$

Among these equations there are $\nu - 1$ independent ones, so that all but one of the mol-numbers may be calculated from the initial numbers. The remaining one can be related to the extent $\Re$. This is the only free variable and it finds its equilibrium value by − once again − the balance of entropy and energy.

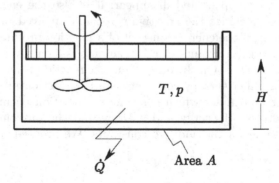

Fig. 12.1 Well-stirred tank under a piston of mass m_p.

To make things simple we require the initial mol-numbers $\tilde{N}_\alpha(0)$ to be such that

$$\frac{\tilde{N}_\alpha(0)}{\gamma_\alpha} \quad \text{is independent of } \alpha \text{ for } \alpha = 1, 2, ...a, \quad \text{and}$$

$$\tilde{N}_\alpha(0) = 0 \quad \text{for } \alpha = a+1, ...\nu.$$
(12.3)

The initial mixture is then called a *stoichiometric mixture* and the conditions ensure that all constituents, which are initially present, may be fully consumed or converted by the reaction.

Also we assume that we start with 1 mol of constituent 1 so that the extent of reaction $\Re$ is bounded by 0 and 1 mol.

12.3 Heat of Reaction

The 1st and 2nd laws of thermodynamics assume the forms

$$\frac{dU}{dt} = \dot{Q} - p\frac{dV}{dt} \quad \text{and} \quad \frac{dS}{dt} \geq \frac{\dot{Q}}{T}$$
(12.4)

under the conditions described in Section 12.1. Since the pressure is constant we may write $(12.4)_1$ as

$$\dot{Q} = \frac{dH}{dt} \quad \text{where} \quad H = U + pV .$$
(12.5)

H is the enthalpy as before. Since, by (12.5), $\dot{Q}$ is converted into H in the isobaric reaction under consideration, H is the relevant value of energy here, cf. Sec. 4.1.

Assuming that the heat of mixing H_{Mix}, cf. (9.1), is zero we may write $H = \sum_{\alpha=1}^{\nu} \tilde{N}_\alpha \tilde{h}_\alpha(T,p)$. We integrate (12.5) between the beginning of the reaction with $\Re = 0$ and the end with $\Re = \Re_F$ and obtain the heat supplied in the reaction between beginning and end

$$Q = \sum_{\alpha=1}^{\nu} \gamma_\alpha \tilde{h}_\alpha(T,p)\Re_F .$$
(12.6)

This heat can be measured, and has been measured for all interesting reactions. It is listed in handbooks by giving the value $\Delta\tilde{h}_R = \sum_{\alpha=1}^{\nu} \gamma_\alpha \tilde{h}_\alpha(T_R,p_R)$ for the reference state $T_R = 298\,\text{K}$ and $p_R = 1\,\text{atm}$. $\Delta\tilde{h}_R$ is called the *heat of reaction*.

We list four values of $\Delta\tilde{h}_R$ for four reactions which we shall consider and discuss later:

$$O \to \tfrac{1}{2}O_2 \qquad\qquad \Delta\tilde{h}_R = -\tilde{h}_R^O + \tfrac{1}{2}\tilde{h}_R^{O_2} \qquad\qquad = -245{,}0\tfrac{kJ}{mol}$$

$$N_2 + 3H_2 \to 2NH_3 \qquad \Delta\tilde{h}_R = -\tilde{h}_R^{N_2} - 3\tilde{h}_R^{H_2} + 2\tilde{h}_R^{NH_3} \qquad = -92{,}4\tfrac{kJ}{mol}$$

$$H_2O + CO_2 \to \tfrac{1}{6}C_6H_{12}O_6 + O_2 \quad \Delta\tilde{h}_R = -\tilde{h}_R^{H_2O} - \tilde{h}_R^{CO_2} + \tfrac{1}{6}\tilde{h}_R^{C_6H_{12}O_6} + \tilde{h}_R^{O_2} = 466{,}3\tfrac{kJ}{mol}$$

$$C + \tfrac{1}{2}O_2 \to CO \qquad\qquad \Delta\tilde{h}_R = -\tilde{h}_R^C - \tfrac{1}{2}\tilde{h}_R^{O_2} + \tilde{h}_R^{CO} \qquad = -110{,}5\tfrac{kJ}{mol}$$

$$(12.7)$$

The first two reactions and the fourth are exothermic, since $\Delta\tilde{h}_R$ is negative, i.e. heat is produced, and the energy decreases, as molecular oxygen, ammonia NH_3 and carbon monoxyd are created. When $\Delta\tilde{h}_R$ is positive, the reaction is endothermic, i.e. heat must be supplied; this is the case for the synthesis of glucose $C_6H_{12}O_6$, cf. $(12.7)_3$.

12.4 Available free energy equals free enthalpy

Elimination of $\dot{Q}$ between the thermodynamic laws (12.4) leads to the inequality

$$\frac{d(U + pV - TS)}{dt} \le 0 . \qquad (12.8)$$

We conclude that the available free energy $\mathfrak{A}$ under these circumstances is equal to the *Gibbs free energy* $U + pV - TS$ which is also sometimes called the *free enthalpy* − particularly in France and Germany. It is universally abbreviated by G. Since G decreases in time, it will be minimal in chemical equilibrium.

If the reactants and the reaction products are all ideal gases we may write the Gibbs free energy as, cf. (9.1), (9.2)

$$G = H \qquad\qquad - T\,S$$

$$= \sum_{\alpha=1}^{\nu} \tilde{N}_\alpha \tilde{h}_\alpha(T) - T \sum_{\alpha=1}^{\nu} \tilde{N}_\alpha \tilde{s}_\alpha(T, p_\alpha)$$

$$= \sum_{\alpha=1}^{\nu} \tilde{N}_\alpha \tilde{h}_\alpha(T) - T \sum_{\alpha=1}^{\nu} \tilde{N}_\alpha \tilde{s}_\alpha(T, p) - T \underline{\sum_{\alpha=1}^{\nu} \tilde{R}\tilde{N}_\alpha \ln \frac{\tilde{N}}{N_\alpha}} ,$$

where the underlined term is the entropy of mixing. $\tilde{R} = Ak = 8.314$ J/mol K is the ideal gas constant, or the molar Boltzmann constant. We have

$$\tilde{h}_\alpha(T) = (z_\alpha + 1)\,\tilde{R}\,(T - T_R) + \tilde{h}_R^\alpha ,$$

$$\tilde{s}_\alpha(T, p) = (z_\alpha + 1)\,\tilde{R}\ln\frac{T}{T_R} - \tilde{R}\ln\frac{p}{p_R} + \tilde{s}_R^\alpha .$$

$$(12.9)$$

Apart from T and p the Gibbs free energy depends on the mol-numbers $\tilde{N}_\alpha$ and hence, by (12.1) on the extent of reaction $\Re$. We make this dependence explicit and write

$$
\begin{aligned}
G(\Re)-G(0) = &\sum_{\alpha=1}^{\nu} \gamma_\alpha \left[\left\{ (z_\alpha+1)\,\tilde{R}\,(T-T_R) + \tilde{h}_R^\alpha \right\} - \right.\\
&\left. -T\left\{ (z_\alpha+1)\,\tilde{R}\ln\tfrac{T}{T_R} - \tilde{R}\ln\tfrac{p}{p_R} + \tilde{s}_R^\alpha \right\} \right]\Re +\\
&+\tilde{R}T \sum_{\alpha=1}^{\nu} \left(\left(\tilde{N}_\alpha(0)+\gamma_\alpha\Re \right) \ln \frac{\tilde{N}_\alpha(0)+\gamma_\alpha\Re}{\sum\limits_{\beta=1}^{\nu}\left(\tilde{N}_\beta(0)+\gamma_\beta\Re\right)} - \tilde{N}_\alpha(0) \ln \frac{\tilde{N}_\alpha(0)}{\sum\limits_{\beta=1}^{\nu}\tilde{N}_\beta(0)} \right).
\end{aligned}
$$

$$(12.10)$$

We postpone the discussion of this formula until we know all its ingredients. What is lacking at this stage is the knowledge of the *entropy of reaction*
$$
\Delta\tilde{s}_R = \sum_{\alpha=1}^{\nu} \gamma_\alpha \tilde{s}_R^\alpha.
$$

12.5 Entropy of Reaction

The constants $\tilde{h}_R^\alpha$ in (12.10) − or rather the combination $\sum\limits_{\alpha=1}^{\nu} \gamma_\alpha \tilde{h}_R^\alpha$ − offer no problem. They are related to the heat of reaction and have been measured, cf. Sect.12.3. The other unknown constant, viz. $\Delta\tilde{s}_R = \sum\limits_{\alpha=1}^{\nu} \gamma_\alpha \tilde{s}_R^\alpha$ may be determined by measuring the mol-numbers $\tilde{N}_\alpha$, or the partial pressures $p_\alpha = \frac{\tilde{N}_\alpha \tilde{R} T}{V}$ in chemical equilibrium. Indeed, the minimum of $G(\Re)$ determines equilibrium and we obtain from (12.10)

$$
\tfrac{1}{\tilde{R}} \sum_{\alpha=1}^{\nu} \gamma_\alpha \tilde{s}_R^\alpha = \underline{\ln \prod_{\alpha=1}^{\nu} \left(\tfrac{p_\alpha}{p_R}\right)^{\gamma_\alpha}} + \tfrac{\Delta\tilde{h}_R}{\tilde{R}T} + \sum_{\alpha=1}^{\nu} \gamma_\alpha\,(z_\alpha+1)\left(1-\tfrac{T_R}{T}-\ln\tfrac{T_R}{T}\right).
$$

$$(12.11)$$

The underlined term has been measured as a function of T and the results have been tabulated for many interesting reactions, e.g. for (12.7)$_1$, the formation of O_2 from O, Table 12.1 shows the result for that reaction.

The heat of reaction $\Delta\tilde{h}_R$ is known from (12.7)$_1$ and so we may use (12.11) to calculate the entropy of reaction $\Delta\tilde{s}_R = \sum\limits_{\alpha=1}^{\nu} \gamma_\alpha \tilde{s}_R^\alpha$ from any line of the table, e.g. the line corresponding to $T = 3000\,\mathrm{K}$. We obtain

$$
O \rightarrow \tfrac{1}{2}O_2 \qquad \Delta\tilde{s}_R = -\tilde{s}_R^O + \tfrac{1}{2}\tilde{s}_R^{O_2} = -56.14\,\frac{\mathrm{J}}{\mathrm{mol\,K}} \qquad (12.12)
$$

Table 12.1 Measured partial pressures in equilibrium as function of temperature for the reaction $O \rightarrow \frac{1}{2}O_2$.

T	$\frac{\sqrt{p_R\,p_{O_2}}}{p_O}$	$\ln \frac{\sqrt{p_R\,p_{O_2}}}{p_O}$
1500	$2{,}113 \cdot 10^5$	12,261
2000	$1{,}259 \cdot 10^3$	7,138
2500	$0{,}563 \cdot 10^2$	4,031
3000	7,495	2,014
3500	1,779	0,576
4000	0,126	−2,072

We conclude that the entropy of reaction is negative. Given that entropy "wants" to grow, this means that entropy favours the atomic constituent.

For other reactions the entropy of reaction is calculated in a similar manner from experimental data. Thus we obtain

$$N_2+3H_2\rightarrow 2NH_3 \qquad \Delta\tilde{s}_R=-\tilde{s}_R^{N_2}-3\tilde{s}_R^{H_2}+2\tilde{s}_R^{NH_3} \qquad =-178.6\tfrac{J}{molK}$$

$$H_2O+CO_2\rightarrow\tfrac{1}{6}C_6H_{12}O_6+O_2 \;\; \Delta\tilde{s}_R=-\tilde{s}_R^{H_2O}-\tilde{s}_R^{CO_2}+\tfrac{1}{6}\tilde{s}_R^{C_6H_{12}O_6}+\tilde{s}_R^{O_2}=-40.1\tfrac{J}{molK}$$

$$C+\tfrac{1}{2}O_2\rightarrow CO \qquad \Delta\tilde{s}_R=-\tilde{s}_R^{C}-\tfrac{1}{2}\tilde{s}_R^{O_2}+\tilde{s}_R^{CO} \qquad =89.14\tfrac{J}{mol\,K}$$

$$(12.13)$$

Among these reactions the formation of carbon monoxyd CO is the only "endo-entropic" one, i.e. it is a reaction with a positive entropy of reaction.

12.6 Energies and entropies in specific chemical reactions

Inspection of the Gibbs free energy (12.10) shows that it consists of three parts:

- two of them are linear in $\Re$, of which the first one, with h_R^α, is the energetic contribution and the second one, with s_R^α, is the entropic contribution,
- the last part is logarithmic in $\Re$; it represents the entropy of mixing.

We proceed to discuss specific examples that provide an insight into the significance of the energetic term and the two entropic ones for the position of chemical equilibrium.

Dissociation and recombination of oxygen

The tableau of graphs in Fig. 12.2 provides an instructive insight into the driving forces of the reaction and the possible shifts of equilibrium.

In all graphs the linear contributions in (12.10) with coefficients

$$\Delta \tilde{h} = \sum_{\alpha=1}^{\nu} \gamma_\alpha \left\{ (z_\alpha + 1)\, \tilde{R}\, (T - T_R) + \tilde{h}_R^\alpha \right\} \quad \text{and}$$

$$\Delta \tilde{s} = \sum_{\alpha=1}^{\nu} \gamma_\alpha \left\{ (z_\alpha + 1)\, \tilde{R} \ln \frac{T}{T_R} - \tilde{R} \ln \frac{p}{p_R} + \tilde{s}_R^\alpha \right\} \qquad (12.14)$$

are the major ones. They have opposite tendencies, since $\Delta \tilde{h}\Re$ has its minimum at $\Re = 1$ mol – on the O_2-side – while $-T\Delta\tilde{s}\Re$ has its minimum at $\Re = 0$ – on the O-side. And at low temperature, here for $T \lesssim 1000\,K$, the energy has the major influence, since $(\Delta \tilde{h} - T\Delta\tilde{s})\Re$ is monotonically decreasing, cf. Fig. 12.2 (a). The entropy of mixing is nearly invisible at low temperatures.

However, when the temperature grows to $T = 3500\,K$, the energetic term is not so predominant anymore. $\Delta \tilde{g} = \Delta \tilde{h} - T\Delta\tilde{s}$ is smaller and the entropy of mixing becomes more noticeable. It is now able to produce a visible minimum of $G(\Re)$ – away from the end-point $\Re = 1$ mol of the $\Re$-interval – at the point $\Re \approx 0.856$ mol. Thus, while there is still a surplus of O_2 in equilibrium, there is also some O present, cf. Fig. 12.2 (b).

At the yet higher temperature $T = 4000\,K$ the entropy has gained the upper hand. The graph $G(\Re)$ has its minimum in the immediate neighbourhood of $\Re = 0$, so that the only constituent in equilibrium is O, cf. Fig. 12.2 (c). We may say that atomic oxygen is "entropically stabilized", if we extrapolate the jargon that was used in the context of phase transitions.

However, the entropy also depends on the pressure p; in fact it decreases with growing p and therefore the entropic stabilization is affected by a high pressure. Fig. 12.2 (d) illustrates that: Comparing that figure with Fig. 12.2 (c) we see that the increase of pressure from 1 atm to 10 atm – with temperature unchanged at $T = 4000\,K$ – has pushed equilibrium back to a predominance of O_2, viz. $\Re = 0.857$ – just where it was for $T = 3500\,K$ and $p = 1$ atm.

We remark that the contribution $-TS_{\text{Mix}}$ of the entropy of mixing is a shallow convex curve with, however, infinite slopes at $\Re = 0$ and $\Re = 1$ mol. This fact guarantees real minima of $G(\Re)$ – rather than the end-point minima of the contribution $(\Delta \tilde{h} - T\Delta\tilde{s})\Re$ that were discussed heretofore. Thus the effect of the entropy of mixing is that the reaction is never complete, no matter how predominant the reactants or the reaction products may be. This is another example for the phenomenon that entropy opposes complete unmixing absolutely. However, for the data of Fig. 12.2 (a) the minimum of $G(\Re)$ lies at

$$1 - \frac{\Re}{1 \text{ mol}} = 1.75 \cdot 10^{-87} (!),$$

which essentially means that among the possible $\frac{1}{2}$ $6.023 \cdot 10^{23}$ O_2-molecules we shall not see a single one dissociated into two O-atoms in a long long time indeed, at least not at $T = 1000\,K$.

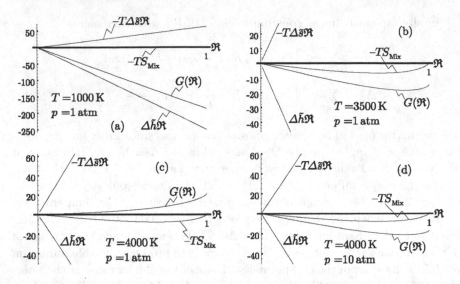

Fig. 12.2 The free enthalpy in kJ for the reaction $O \rightarrow \frac{1}{2}O_2$
as a function of the extent or reaction $\Re$ for different
temperatures and pressures.

Haber-Bosch synthesis

The synthesis of ammonia, which the world craves for the production of fertilizers and explosives, is a non-trivial paradigm for the mutual competition of energy and entropy in a chemical reaction. Let us consider this:

The stoichiometric formula and essential data for the synthesis of ammonia are given by $(12.7)_2$ and $(12.13)_1$. For the reference temperature $T_R = 298\,\mathrm{K}$ and reference pressure $p_R = 1\,\mathrm{atm}$ the energetic linear term $\Delta \tilde{h}_R \cdot \Re$ favours ammonia and the entropic term $-T \Delta \tilde{s}_R \Re$ favours the constituents nitrogen and hydrogen; but the energetic contribution is stronger. The entropy of mixing is not totally negligible. Indeed, the graph of $G(\Re)$ is slightly curved due to the entropy of mixing, cf. Fig. 12.3, and it has a minimum at $\Re \approx 0.98\,\mathrm{mol}$.

Therefore at reference conditions the mixture should be nearly entirely ammonia. Unfortunately the situation is more complex. Indeed, the curve for $T = T_R$ and $p = p_R$ in Fig. 12.3 is totally irrelevant, because N_2 and H_2 do not react at all before being split into their atomic variants. For that purpose we need a catalyser – e.g. an iron sheet. But the catalyser works well only at $T = 500°C$. At that temperature, however, the entropic contribution $-T \Delta \tilde{s}$ – having a minimum at $\Re = 0$ – dominates the energetic one $\Delta \tilde{h}$, so that there is still no ammonia, cf. Fig. 12.3 and the curve marked $T = 798\,\mathrm{K}$, $p = 1\,\mathrm{atm}$.

In order to obtain ammonia at $500°C$ we need to decrease the entropic term $\Delta \tilde{s}(T, p)\Re$ and that may be done by increasing the pressure, cf. (12.10).

Figure 12.3 shows a graph marked $T = 798\,\mathrm{K}$, $p = 200\,\mathrm{atm}$ and that graph has a minimum at about $\Re = 0.42\,\mathrm{mol}$, so that a sizable quantity of ammonia is produced.

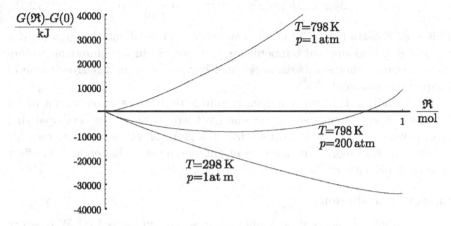

Fig. 12.3 Free enthalpies for the ammonia synthesis

12.7 Photosynthesis

The dilemma of photosynthesis

Plants produce glucose $C_6H_{12}O_6$ in their leaves from the carbon dioxide of the air and from water of the soil. They release oxygen. The process is called photosynthesis because it proceeds with the cooperation of the light from the sun. Not all steps of the reaction seem to be fully understood by the biochemists but the overall stoichiometric equation is clear. It reads

$$CO_2 + H_2O \rightarrow \tfrac{1}{6}C_6H_{12}O_6 + O_2 \ . \tag{12.15}$$

The heat of reaction $\Delta\tilde{h}_R$ and the entropy of reaction $\Delta\tilde{s}_R$ have been measured and they have values listed in handboooks of chemistry, viz.

$$\Delta\tilde{h}_R = 466.3 \ \frac{\mathrm{kJ}}{\mathrm{mol}} \quad \text{and} \quad \Delta\tilde{s}_R = -40.1 \ \frac{\mathrm{J}}{\mathrm{mol\,K}} \ . \tag{12.16}$$

Therefore the reaction is endothermic and "exentropic", the energy grows and the entropy decreases. That is the worst combination that can occur, because the first law requires that we supply heat and the second law demands a release of heat. Indeed, if the reaction occurs at $T = T_R = 298\,\mathrm{K}$ and $p = p_R = 1\,\mathrm{atm}$, the molar heat of reaction must obey

$$\tilde{q} = \Delta\tilde{h}_R > 0 \quad \text{and} \quad \tilde{q} \leq T_R\Delta\tilde{s}_R < 0 \ . \tag{12.17}$$

These conflicting predictions of the thermodynamic laws may be summarized by calculating the molar free enthalpy of reaction which is positive, namely

$$\Delta \tilde{g}_R = \Delta \tilde{h}_R - T_R \Delta \tilde{s}_R = 478.3 \ \frac{kJ}{mol} > 0 \ . \tag{12.18}$$

This sign contradicts (12.8). Nor can we do much about the growth of g by changing pressure and temperature — as we did in the ammonia synthesis, — because photosynthesis works at 1 bar and only in a narrow range of temperatures around 25°C.

So, there is a dilemma! We must conclude that the reaction cannot occur by itself. Apart from a supply of energy there must be an accompanying process which increases the entropy far enough to offset the negative entropy of reaction. And more: The increase of entropy must be big enough to effect an overall decrease of the free enthalpy.

Energy by radiation

The supply of energy is no problem, since the sun sends 1341 W to every square meter on the earth which is perpendicular to the incoming radiation. 75% of that power reaches the surface of the earth and the plant leaves absorb 65% of what is left, primarily the red-yellow component of the spectrum, which is why the leaves appear green. Thus we may estimate that a leaf absorbs 650 W/m². At the same time the leaf at temperature T radiates off an energy flux that — by the law of Stefan-Boltzmann — is proportional to T^4. The factor of proportionality is $\sigma = 5.67 \cdot 10^{-8} \frac{W}{m^2 K^4}$ which is known as the Stefan constant. [We assume here that the emission coefficient equals 1. In reality it is smaller but it should be close to 1.][1]

Plant physiologists have found out that a reasonable value of plant growth is given by 1 gr, or $\frac{1}{180}$ mol glucose per hour and square meter of leaf. By (12.15) this requires a reaction rate of $\frac{1}{30} \frac{mol}{hm^2}$. For simplicity we base our calculation on a 30 hour growth which produces $\frac{1}{6} \frac{mol}{m^2}$ glucose and requires $1 \frac{mol}{m^2}$ water and $1 \frac{mol}{m^2} CO_2$. The required energy is, cf. (12.16)₂

$$466.3 \ \frac{kJ}{30hm^2} = 4.3 \ \frac{J}{sm^2} \ .$$

Taken altogether the balance of radiative energies and of the heat of reaction thus reads

$$650 \frac{W}{m^2} - \sigma T^4 = 4.3 \frac{J}{sm^2} \ . \tag{12.19}$$

Hence follows $T = 327\,K$, and that is much too hot for photosynthesis to proceed. Also, on a more intuitive level, we know that this is surely not the temperature of a leaf on a living plant. So the question arises how the leaf stays cool?

---

[1] For a brief account of the thermodynamics of radiation see Chaps. 18, 19.

Why a plant needs excess water

The key to the answer is evaporation of water. Indeed, farmers, gardeners and housewives know that a plant consumes more water – much more – than the $1 \frac{mol}{m^2}$ in 30h which the stoichiometric formula (12.15) demands: The actual demand may be one hundred to one thousand times as much (!), or, generically, x times as much. The plant absorbs all that water in the roots, passes it upwards to the leaves and evaporates it there. Thus the plant cools its leaves much in the same way as animals cool their skins, by evaporation. Let us calculate the water flux which is needed to maintain the temperature $T = 298\,\text{K}$.

The molar heat of evaporation $\tilde{r}_{H_2O} = 43.9\,\frac{kJ}{mol}$ of the water flux $x \cdot \frac{mol}{30hm^2}$ must be added to the energy balance (12.19) which then reads

$$\left(650\frac{W}{m^2} - \sigma T^4\right) = 4.3\,\frac{J}{sm^2} + x\,\tilde{r}_{H_2O}\frac{mol}{30hm^2}\,. \qquad (12.20)$$

For $T = 298\,\text{K}$ this equation implies

$$x \approx 490, \quad \text{hence water consumption in 30h: } J_W = 490\,\frac{mol}{m^2}\,. \qquad (12.21)$$

Compared to the flux of water demanded by the stoichiometric equation for the stipulated growth rate of glucose, the cooling water flux is thus 490 times bigger (!). Nearly all the water evaporates and only a small fraction is used to form glucose.

Why a plant needs excess air

The evaporation of water cannot help, however, with the wrong sign of $\Delta\tilde{g}_R$ in (12.18), because $\tilde{g}$ does not change when liquid water is converted to vapour, cf. Sect.10.3. We are therefore still looking for an accompanying process that increases entropy to such an extent that g decreases. We propose that the mixing of the evaporated water with the surrounding or passing air may serve as such a process, because it creates a considerable amount of entropy of mixing. Let us consider that proposition:

The CO_2 used for the reaction (12.15) has a partial pressure $p_{CO_2} = 3 \cdot 10^{-4}$ bar in air. This means that the stoichiometric equation (12.15) requires an air flux

$$J_{Air} = \tfrac{1}{3}10^4\frac{mol}{m^2} \qquad (12.22)$$

at least, if we wish to obtain the stipulated plant growth of $\frac{1}{6}\frac{mol}{m^2}$ glucose in 30h.

The air flux (12.22) is incapable of carrying off the evaporated water flux of $489\frac{mol}{m^2}$, cf. (12.21); at least it cannot do that at a temperature of 298 K without becoming foggy. Since we see no fog appearing when a plant grows, we must at least provide about 4.5 times as much air. Indeed, air becomes saturated when it contains 2 mass % of water.[2]

However, we shall need more air than (12.22), not only to avoid fog but also to provide the water vapour with the possibility to mix with an amount of air large enough to satisfy the thermodynamic requirement that the free enthalpy must decrease. Let us denote the necessary air flux generically by $y \cdot J_{\mathrm{Air}}$ and let us proceed to determine the excess factor y.

We consider the glucose reaction in a system of plant, air and water consisting of

- initially: $y\frac{1}{3}10^4$ mol air, x mol liquid water and a plant
- finally: $y\frac{1}{3}10^4$ mol air[3], $x - 1$ mol water vapour and with the plant grown by $\frac{1}{6}$ mol $C_6H_{12}O_6$.

The system is under the fixed pressure $p = p_R$ and has the constant temperature $T = T_R$. From beginning to end the reaction will take 30 h of efficient photosynthesis as we have explained before.

The partial pressure $p_{H_2O}^f$ of the water vapour at the end of the reaction is given by

$$p_{H_2O}^f = p_R \frac{x-1}{y\frac{1}{3}10^4 + (x-1)}. \tag{12.23}$$

and the molar enthalpy and molar entropy of the evaporated water are

$$\tilde{h}_{H_2O}^f = \tilde{h}_{H_2O}^i + \tilde{r}_{H_2O}(T_R) \quad \text{and} \quad \tilde{s}_{H_2O}^f = \tilde{s}_{H_2O}^i + \frac{\tilde{r}_{H_2O}(T_R)}{T_R} - \tilde{R}\ln\frac{p_{H_2O}^f}{p_{H_2O}(T_R)}, \tag{12.24}$$

where $p_{H_2O}(T_R)$ is the saturation pressure of water at $T_R = 298\,K$ and it amounts to $p_{H_2O}(T_R) = 0.0317\,bar$.

By (12.23), (12.24) the decrease of molar free enthalpy of the water during the reaction is thus given by

$$\Delta\tilde{g}_{H_2O} = \tilde{R}T_R\ln\left[\frac{p_R}{p_{H_2O}(T_R)}\frac{x-1}{y\frac{1}{3}10^4 + (x-1)}\right]. \tag{12.25}$$

This term must be added to the $\Delta\tilde{g}_R$ of the stoichiometric reaction in order to account for the evaporation of water.

Therefore we obtain for the total change of molar free enthalpy

[2] This follows from the interesting thermodynamic theory of moist air which, however, we do not discuss in this book.
[3] In the final state the air has been stripped of part of its initial CO_2 and it has gained some O_2. We ignore such tiny effects because they do not affect the main conclusion of this section.

$$\Delta\tilde{g}_R + \Delta\tilde{g}_{H_2O} = 478.3\frac{kJ}{mol} + \tilde{R}T_R \ln\left[\frac{p_R}{p_{H_2O}(T_R)}\,\frac{x-1}{y\frac{1}{3}10^4 + (x-1)}\right]. \quad (12.26)$$

We determine the excess factor y of air from the criterion that the free enthalpy must decrease, cf. (12.8). A plot of the graph of the function on the right hand side of (12.26) shows that it falls below zero at $y \approx 9$.

Therefore we conclude that the plant needs approximately 9 times as much air as would be required for the CO_2 supply of the stoichiometric formula (12.15). The excess air is needed to allow the "cooling water" to evaporate and increase its entropy by mixing. Once again we have a system here, which allows one component − the plant − to lower its entropy when another component − the water − gains enough entropy to compensate for the loss.

Discussion

The dilemma of photosynthesis has not escaped the attention of physicists and chemists altogether, although plant physiologists rarely mention it in their detailed discussion of the photosynthetic process in C3 and C4 plants. One well-publicized mention of the dilemma occurs in the book "What is life ?" by E. Schrödinger. He seems to say that the plant sets its free enthalpy balance right by the use of solar radiation.

That may be so. If it is, we do not understand the specifics of the process. In any case, our treatment of plant growth with supplies of excess water and air provides an alternative.

And yet, there remains a problem: How do water plants balance their free enthalpy? It can hardly be expected that they evaporate water and mix it with air. But then, water plants do not need cooling; they are cooled by the surrounding water. We do not have enough knowledge about the physiology of water plants to apply thermodynamics to it.

13

Shape memory alloys

The rich thermomechanical properties of shape memory alloys result from an austenitic-martensitic phase transition of the metallic lattice and from a twinning deformation in the low temperature phase, the martensite.

The thermodynamicist is interested in the phenomena, because — once again like in all phase transitions — they provide a non-trivial example for the competition between energy and entropy, with entropy gaining the upper hand at high temperatures. Recognizing this we are able to produce a model for shape memory behaviour which is capable of simulating all the observed features, at least qualitatively. Actually we describe several models, a thermodynamic one, a kinetic one and a numerical one. The latter employs molecular dynamics; it is particularly instructive and the accompanying CD permits the reader to view the crystalline rearrangement by observing the atoms shifting between their potential wells.

Shape memory alloys have found numerous small technical applications, and applications in the medical field. We start the chapter with a selective review.

13.1 Phenomena and applications

Upon heating, some metallic alloys like NiTi, CuZnAl, CuAlNi restore their original shape after a plastic deformation. We say that the alloy "remembers" the original shape and call such materials *shape memory alloys*. Figure 13.1 illustrates the memory effect schematically.

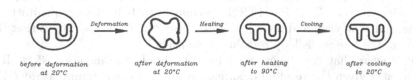

| before deformation | after deformation | after heating | after cooling |
| at 20°C | at 20°C | to 90°C | to 20°C |

Fig. 13.1. Memory effect in a NiTi wire. The wire forms the logo-gram of our institution, the Technical University Berlin.

The memory effect is a side effect of the strong temperature dependence of the load-deformation diagrams of such alloys. Figure 13.2 shows four such diagrams for increasing temperatures T_1 through T_4.

At the low temperature T_1 the load-deformation curve is much like that of a plastic body: A loading-unloading experiment with a small tensile load leads to elastic deformations along a straight line through the origin. However, as soon as the load reaches a critical value, the yield load, the deformation increases without further increase in load and we say that the loaded specimen *yields* like a plastic body. In contrast to a plastic body, however, the yield comes to an end on a second elastic line along which the load may increase far beyond the yield load. Subsequent unloading leads to the residual deformation D_1, cf. Fig.13.2.

For the somewhat higher temperature T_2 this behaviour is qualitatively unchanged except that the yield load is a little smaller. This low-temperature behaviour is called *quasiplasticity*.

Under compressive loads one observes an analogous behaviour so that a hysteresis loop surrounds the origin — always at low temperature — in an alternating tensile and compressive loading experiment.

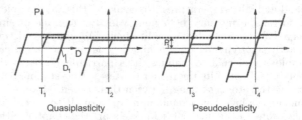

Fig. 13.2. Load-deformation diagrams for different temperatures
Quasiplasticity and pseudoelasticity

At the higher temperature T_3 we observe a different behaviour altogether. To be sure there is still an elastic line through the origin and a yield line and a second elastic line. But, upon unloading, when the load falls below the *recovery load*, the specimen recovers the deformation that it yielded before and moves back to the original elastic line. It then returns to the origin when the unloading is completed. This behaviour is called *pseudoelastic*; it is elastic, because the specimen returns to the origin, but it is only *pseudo*elastic, because there is a hysteresis loop in the loading-unloading cycle. When the temperature is increased to T_4, cf. Fig. 13.2 the pseudoelastic behaviour is qualitatively unchanged. The yield and recovery loads increase at the same rate so that the height of the hysteresis loop remains unchanged.

It is clear that the diagrams of Fig. 13.2 imply the memory effect: If a specimen is given the residual deformation D_1 at the low temperature T_1, it must return to the origin at $D = 0$ upon heating to T_4 since that is the only load-free state at high temperature.

Typically the temperature interval $T_4 - T_1$ for the diagrams of Fig. 13.2 is 40 K and the mean temperature may be at room temperature. The maximal recoverable deformation is 6 to 8 %.

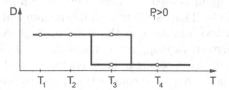

Fig. 13.3. Deformation-temperature diagram for a constant tensile load

If one measures load-deformation isotherms for many temperatures, one may construct deformation-temperature diagrams for fixed loads. Figure 13.3 shows such a diagram for a positive load P_1 in schematic form. It is indicated in that figure how the points of the (D, T)-diagram result from the curves of the (P, D)-diagrams of Fig. 13.2. We conclude that the (D, T)-hysteresis reflects the (P, D)-hystereses.

Inspection of the (D, T)-diagram shows that a memory alloy may be said to store two deformations in its memory: a small one for high temperature and a large one for low temperature. If the specimen is subjected to an alternating temperature, it will alternately expand and contract. Therefore memory alloys may be used to build temperature-controlled actuators. Or else, if we install a suitable mechanical power transmission, one may construct a heat engine. Figure 13.4 shows two interesting − and functioning − constructions.

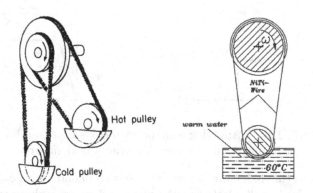

Fig. 13.4. Two memory-engines
Left: Engine of A.D. Johnson
Right: Engine of F.E. Wang

Often repeated loading and unloading may induce an internal stress field which forces the − externally unloaded − specimen to alternate between two deformations for alternating temperatures. This phenomenon is called the *two-way shape memory* effect.

An interesting application occurs in the medical field; it is illustrated in Fig. 13.5$_{\text{left}}$. The figure shows a broken jaw bone which must be fixed by a splint. In the application of the splint it is important to ensure that the bone-ends are tightly pressed against each other. That requires much skill of the surgeon when a steel splint is used and even then the compressive force is still smaller than is desirable. That is where a shape memory alloy can help. The memory splint is screwed into place in the state of large deformation on the upper border of the hysteresis loop, and − naturally − at 36°C. The implant is then heated to 42°C (say) whereupon it contracts and pushes the bone ends firmly against each other. The contraction persists even after the splint assumes body temperature again, because now the state of the material lies on the lower border of the hysteresis loop. The relevant states are marked by circles in Fig. 13.5$_{\text{left}}$.

Another medical application occurs in dentistry when braces are used for readjusting the position of teeth. The brace is pseudoelastic and prede-formed and prestressed so as to be on the recovery line upon installation, cf. Fig. 13.5$_{\text{right}}$. The teeth move under the load and the deformation D decreases; as it does, the load P is unchanged for some considerable period of contraction. Therefore a readjustment of the brace is not necessary for a long while.

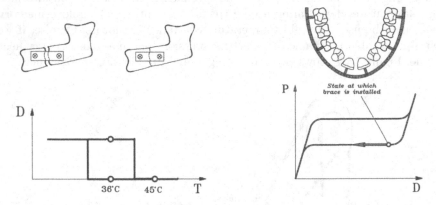

Fig. 13.5. Left: Applying a shape-memory splint to a broken jaw
Right: Shape memory for braces in dentistry

The key to the understanding of the observed phenomena lies in the observation that the metallic lattice of the alloy undergoes a phase transition. At high temperatures the highly symmetric austenitic phase prevails, whereas at low temperatures the lattice is in the less symmetric martensitic phase. And martensite may form twins. In the virginal state at low temperature different twins are ideally present in equal proportions. By uniaxial loading, *one* of the twins is favoured by the direction of the load. That twin will then form at the expense of the others in a process called *twinning*. At high temperatures no martensite occurs in an unloaded specimen. But the application of a load may force the prevailing austenite into a martensitic twin variant. That happens on the pseudoelastic yield line. On the recovery line the reverse transition occurs.

On the basis of these considerations we may – in imagination – construct a model that is capable of simulating the observed phenomena.

13.2 A model for shape memory alloys

The basic element of the model is what we call a lattice particle, a small piece of the metallic lattice. Fig. 13.6left shows this lattice particle in three equilibrium configurations. They are denoted by A and $M_\pm$ which stands for austenite and two martensitic twins respectively.

One may imagine that the martensitic twins are sheared versions of the austenitic particle. The postulated form of the potential energy as a function of the shear length Δ is characterized by stable lateral minima which correspond to the martensitic twins and by a metastable central minimum for the austenitic particle. In-between those minima there are energetic barriers, cf. Fig. 13.6. For simplicity we represent this three-well-potential by a train of three parabolae, viz.

$$\Phi(\Delta) = \begin{bmatrix} \Phi_A = \Phi_0 + K_A \Delta^2 & |\Delta| \le \Delta_s \\ & \text{for} \\ \Phi_{M_\pm} = K_M(\Delta \mp J)^2 & |\Delta| > \Delta_s \end{bmatrix} . \qquad (13.1)$$

It is important for the proper simulation of the observed phenomena to choose $\Phi_0 > 0$ and $K_M > K_A$ so that the lateral potential wells are narrower and deeper than the central one.

If a load P is exerted on the lattice particles, cf. Fig. 13.7right, its potential energy must be added to $\Phi(\Delta)$. The potential energy of the load is proportional to the shear length Δ with P as the factor of proportionality. Thus the effective potential energy is deformed as shown in Fig. 13.6right.

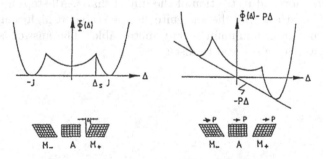

Fig. 13.6. Lattice particle and its potential energy
Left: unloaded particle
Right: particle under a load P

A model for a tensile specimen results by assembling lattice particles into layers and forming stacks of layers — as indicated in Fig. 13.7 — oriented at 45°C to the direction of the tensile force. On the left hand side of the figure the specimen with alternating layers of $M_\pm$ particles corresponds to the unloaded specimen at low temperature. We proceed to discuss the comportment of the model under vertical loading.

If the tensile force is small, the layers are subjected to a small shear force which makes the M_+-layers flatter and the M_--layers steeper. The vertical components of the shear lengths Δ_i of all N layers add up to the deformation D of the specimen

$$D = L - L_0 = \frac{1}{\sqrt{2}} \sum_{i=1}^{N} \Delta_i \ . \tag{13.2}$$

Upon unloading the layers fall back to their original positions which means that the deformation was elastic. If, however, the load is increased, there comes the point when the M_--layers flip over into the M_+-position and thus increase their shear length drastically. Consequently the deformation grows drastically at the load where flipping occurs. Subsequent unloading does not induce the layers to flip back. Rather they will all move to the unloaded M_+-position and therefore a large residual deformation results.

In this plausible and suggestive manner we understand the initial elastic deformation, the yield as a consequence of flipping, and the residual deformation — all at low temperature.

Upon heating the austenitic phase is formed. Consequently the layers straight-en up and the specimen will shorten as shown on the right hand side of Fig. 13.7. For the naked eye the specimen has then already recovered its original shape. To be sure, it still has a different internal structure and a smooth surface, but that can only be detected under the microscope. Subsequent cooling reproduces the martensitic twins and, in all probability, equal numbers of M_+ and M_-. Therefore the body may return to its old state, both with respect to shape and internal structure.

Thus we have led the model specimen through the full cycle of elastic-plastic deformation and restitution of the initial shape. All steps are easy except one: We must ask why the austenite can be stable at high temperatures although its potential minimum is only metastable? The answer is entropic stabilization which we proceed to discuss.

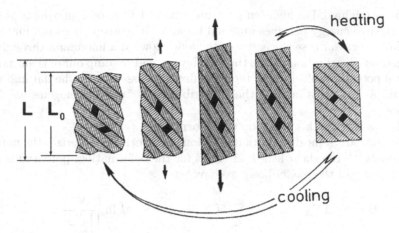

Fig. 13.7. A model for a tensile specimen

13.3 Entropic stabilization

Why is the austenite stable at high temperature when martensite is energetically more favourable? It is true that the martensitic minima are lower but they are also narrower — recall $K_M > K_A$ — and that feature stabilizes the austenite for high temperatures. Let us consider:

The lattice layers fluctuate about their minima and that motion is more lively for a higher temperature. Thus occasionally a layer will be able to surmount the barriers. It is more easy to pass the small $A \rightarrow M$ barrier than the large $M \rightarrow A$ barrier. Thus we should expect martensite to be dominant. *However*: Since the M-wells are narrower than the A-well, the M-layers hit the barrier more often than the A-layers hit theirs; and each time they have the chance to overcome the barrier. This advantage may compensate — for the M-layers — the disadvantage of the deeper well so that the layers tend to assemble in the A-well.

This is a case of entropic stabilization which we proceed to discuss in terms of the competition of energy and entropy.

Energy "tries" to become minimal by assembling the layers in the depth of their potential wells and entropy "tries" to become maximal by distributing the layers evenly over the available shear lengths. In this situation it is the free energy

$$F = U - TS \tag{13.3}$$

that in fact becomes minimal. For low temperatures the second term may be neglected and the free energy assumes a minimum, because the energy becomes minimal. For high temperatures the first term of F may be neglected and the free energy becomes minimal because the entropy becomes maximal.

In the present case the entropic tendency toward a maximum through homogeneous distribution means that the layers tend to jump out of their narrow lateral potential wells at high temperatures and assemble in the flat and wide central well. In this manner their distribution is "more homogeneous" than before.

Let us discuss this situation more formally:

We introduce the distribution function N_Δ which characterizes the number of layers with the shear length Δ. Thus for the deformation, potential energy and entropy of the model body we have[1]

$$D = \sum_\Delta \Delta N_\Delta , \quad U = \sum_\Delta \Phi(\Delta) N_\Delta , \quad S = k \ln \frac{N!}{\prod_\Delta N_\Delta!} , \quad (13.4)$$

where the summation extends over all possible shear lengths. Deformation, potential energy and entropy of the phases A and $M_\pm$ are similarly defined except that the Δ's run over Δ_A, i.e. $|\Delta| \leq \Delta_s$, and $\Delta_{M_\pm}$, i.e. $\Delta \gtrless \pm \Delta_s$ respectively. We use the Stirling formula and write the free energies, numbers of layers and deformations of each phase as

$$F_A = \sum_{\Delta_A} \left(\Phi_A(\Delta) + kT \ln \frac{N_\Delta}{N_A} \right) N_\Delta, \quad N_A = \sum_{\Delta_A} N_\Delta, \quad D_A = \sum_{\Delta_A} \Delta N_\Delta$$

$$(13.5)$$

$$F_{M_\pm} = \sum_{\Delta_{M_\pm}} \left(\Phi_{M_\pm}(\Delta) + kT \ln \frac{N_\Delta}{N_{M_\pm}} \right) N_\Delta, \quad N_{M_\pm} = \sum_{\Delta_{M_\pm}} N_\Delta, \quad D_{M_\pm} = \sum_{\Delta_{M_\pm}} \Delta N_\Delta.$$

In order to find the equilibrium values of the numbers of layers in the phases A and $M_\pm$ we minimize F_A and $F_{M_\pm}$ under constraints which fix N_A, D_A and $N_{M_\pm}$, $D_{M_\pm}$ respectively. Thus we obtain

$$N_\Delta = N_A \frac{e^{-\frac{\Phi(\Delta) - \beta_A \Delta}{kT}}}{\sum_{\Delta_A} e^{-\frac{\Phi(\Delta) - \beta_A \Delta}{kT}}} \quad \text{and} \quad N_\Delta = N_{M_\pm} \frac{e^{-\frac{\Phi(\Delta) - \beta_{M_\pm} \Delta}{kT}}}{\sum_{\Delta_{M_\pm}} e^{-\frac{\Phi(\Delta) - \beta_{M_\pm} \Delta}{kT}}}$$

$$(13.6)$$

for $|\Delta| \leq \Delta_s$ and $\Delta \gtrless \pm \Delta_s$ respectively. The β's are Lagrange multipliers that take care of the constraints of fixed deformations. Their values follow from the constraints. In order to exploit these we need to convert the sums into integrals. This is done by assuming that the number of shear lengths between

---

[1] In (13.4)$_1$ we incorporate the factor $\sqrt{2}$ from (13.2) into the shear length. Thus we simplify the notation.

$\varDelta$ and $\varDelta + d\varDelta$ is equal to $Yd\varDelta$, where Y is some factor of proportionality. We simplify the integrations by carrying them out between $-\infty$ and $+\infty$ for all wells instead of only about the proper ranges of the parabolic wells; this procedure may be justified, if the layers are all situated near the minima of the wells. Thus we obtain from $(13.5)_3$

$$\beta_A = 2K_A \frac{D_A}{N_A} \qquad \text{and} \qquad \beta_{M_\pm} = 2K_M \left(\frac{D_{M_\pm}}{N_{M_\pm}} \mp J \right) . \qquad (13.7)$$

Insertion into (13.6) provides the distribution functions in explicit form and when these are used in $(13.5)_1$ we obtain

$$\frac{F_A}{N_A} = \underline{\varPhi_0 + K_A \left(\frac{D_A}{N_A} \right)^2} + \frac{1}{2}kT \ln K_A - kT \ln \left(Y\sqrt{\pi kT} \right)$$

$$\tag{13.8}$$

$$\frac{F_{M_\pm}}{N_{M_\pm}} = \underline{K_M \left(\frac{D_{M_\pm}}{N_{M_\pm}} \mp J \right)^2} + \frac{1}{2}kT \ln K_M - kT \ln \left(Y\sqrt{\pi kT} \right) .$$

From basic thermodynamics we recall that $P = \frac{\partial F}{\partial D}$ is the load. Therefore, by (13.7) and (13.8), we conclude that β_A and $\beta_{M_\pm}$ are the loads on the phases. Those loads are needed to maintain the deformations D_A and $D_{M_\pm}$.

We recognize that for $T = 0$ in (13.8) we retain only the underlined parts. As functions of D/N – the contribution of one layer to the deformation – these parts are equal to the potential energy functions (13.1). With increasing temperature the functions F_A and $F_{M_\pm}$ are shifted vertically at different rates: $F_{M_\pm}$ more so than F_A because $K_M > K_A$ holds. Thus with increasing temperature we obtain the graphs of Fig. 13.8, where $F(0)$ has arbitrarily – and without loss of generality – been set equal to zero in all cases.

Inspection shows that for small temperatures the martensitic twins are stable, i.e. that they have the lowest minima of the free energy. For some higher temperature all minima are equally deep, but then for high temperatures the austenitic minimum is deepest. The phase change $M_\pm \iff A$ occurs when all minima have the same depth. Here we have the analytic – and graphical – version of the previously discussed entropic stabilization. Indeed, the graphs of Fig. 13.8 differ only by the entropic, T-dependent contributions, cf. (13.8).

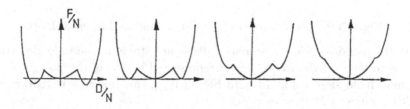

Fig. 13.8. Free energies of the phases A and $M_\pm$ as functions of D/N for increasing temperature, cf. (13.8)

For the benefit of those readers who know the Landau-Devonshire theory of phase transitions we mention that the sequence of graphs of Fig. 13.8 qualitatively resembles free energies of that theory. To be sure, in Landau's case each graph is represented by a single temperature-dependent six-order polynomial, but the relative positions of the minima and the non-convex character of the free energy are essentially the same.

13.4 Pseudoelasticity

Neither here nor in the Landau-Devonshire model do hystereses appear. Therefore the model is not yet capable of describing the load-deformation curves of Fig. 13.2. We suggest that the proper representation of the hysteresis loops requires that a penalty for the formation of phase interfaces be taken into account. That is the idea which we persue in this section, albeit not for the full spectrum of possibilities represented in Fig. 13.2 but only for pseudoelasticity.

The pseudoelastic hysteresis is richer than we have discussed sofar. Indeed, when one interrupts the yield process by unloading or the recovery process by reloading, one observes the behaviours shown in Fig. 13.9. The steep lines inside the hysteresis loop are elastic, i.e. reversible. However, upon unloading, and when the load falls below the diagonal line − from the upper left to the lower right corner of the loop −, recovery occurs inside the hysteresis loop at constant load. Analogously one can observe internal yield, if the diagonal line is approached from below. By conducting a loading-unloading process properly, one may thus trace out inner loops, cf. Fig. 13.9$_{right}$.

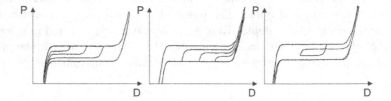

Fig. 13.9. Internal recovery, internal yield and internal loop

In the pseudoelastic range martensite is not stable in the unloaded state. That is to say that the free energy is represented by one of the two graphs on the right hand side of Fig. 13.8. In Fig. 13.10$_{top}$ this function is redrawn for positive values of D.

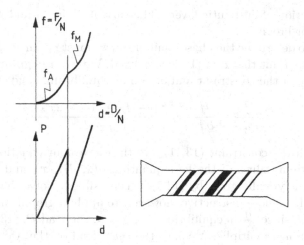

Fig. 13.10. Left: Specific free energy and load as functions
of specific deformation
Right: Schematic view of a tensile specimen
during pseudoelastic yielding

The lower parabola corresponds to austenite, the upper one to martensite. If both phases are present, as they are in the pseudoelastic range, the
tensile specimen alternates between austenitic and martensitic stripes, cf.
Fig. 13.10$_{\text{right}}$. Hundreds or thousands of such stripes can occur and we assume
that their number may be set proportional to $N_A \frac{N_M}{N}$, where N_A and N_M are
the number of layers in the austenitic and martensitic phase respectively and
$N = N_A + N_M$. This expression is plausible, since it represents the expectation value of the number of interfaces, if their arrangement is random. Each
interface is endowed with a certain energetic penalty A, because the lattice
must be deformed for the formation of an interface in a coherent lattice. We
call A the coherency factor and write the total energy penalty in the form

$$A N_A \frac{N_M}{N} \quad \text{with} \quad A > 0 \sim \text{coherency factor .} \tag{13.9}$$

Thus the free energy and the deformation of the specimen are given by

$$F = F_A + F_M + A N_A \frac{N_M}{N} \quad \text{and} \quad D = D_A + D_M, \tag{13.10}$$

and for the specific values $f = F/N$ and $d = D/N$, referred to a layer, we
have

$$f = x_A f_A(d_A) + (1 - x_A) f_M(d_M) + A x_A(1 - x_A)$$

$$d = x_A d_A + (1 - x_A) d_M . \tag{13.11}$$

x_A is the fraction of austenitic layers. The variable T in f_A and f_M has been suppressed for brevity.

In order to determine the phase equilibrium we must minimize f in $(13.11)_1$ under the constraint that d in $(13.11)_2$ is fixed. We use a Lagrange multiplier λ to take care of the constraint and obtain as equilibrium conditions

$$\lambda = \frac{\partial f_A}{\partial d_A} = \frac{\partial f_M}{\partial d_M} = \frac{f_M - f_A + A(1 - 2x_A)}{d_M - d_A} . \tag{13.12}$$

Together with the constraint $(13.11)_2$ we thus have four equations to determine equilibrium values of the four quantities d_A, d_M, x_A and λ. Since f_A and f_M are analytically given by (13.5), it is possible to solve these equations analytically. It is more instructive, however, to apply a graphical procedure.

First of all, since $\frac{\partial f}{\partial d}$ in equilibrium is equal to the load P, $(13.12)_{1,2}$ show that the Lagrange multiplier λ equals the *equal* load on the two phases. The load as the derivative of the free energy is shown in Fig.13.10. In the model it is represented by a straight graph in both phases.

For $\lambda = P$ and $f = \int P(d)\mathrm{d}d$ $(13.12)_3$ implies

$$P(d_M - d_A) - \int_{d_A}^{d_M} P(d)\mathrm{d}d = -A(1 - 2x_A) . \tag{13.13}$$

This is the condition for phase equilibrium and we shall use it to determine the load and the corresponding equilibrium deformations d_A and d_M. (13.13) may be used for a graphical construction of these values. They will depend on x_A.

Let us first consider $x_A \lesssim 1$, i.e. little martensite in equilibrium with the predominant austenite. In that case the right hand side of (13.13) is equal to A and the rectangle $P(d_M - d_A)$ must be chosen so as to be bigger by the amount A than the integral $\int_{d_A}^{d_M} P(d)\mathrm{d}d$ under the graph $P(d)$. Figure 13.11$_{\text{left}}$ shows how the corresponding values P^0, d_M^0, d_A^0 may be found: obviously the triangular areas I and II must differ by A. Next let $x_A \gtrsim 0$ hold, i.e. little austenite coexists with the prevailing martensite. The right hand side of (13.13) equals $-A$ in that case and the second graph of Fig.13.11 shows how the corresponding values P^1, d_M^1, d_A^1 may be found: the triangular areas III and IV must differ by A.

It stands to reason[2] that the loads P^0 and P^1, which characterize the beginning phase transitions, represent the observed yield load and recovery load respectively. If this is the case, the hysteresis loop has the area $2A$ as indicated in Fig.13.11$_{\text{right}}$.

---

[2] Not more than that! We should like to *prove* that the yield and recovery loads may be calculated in this way.

The equilibrium load P^x that characterizes a phase equilibrium with x_A different from 0 and 1 follows from (13.13) in much the same manner as P^0 and P^1 and, once found, the corresponding values d_A^x, d_M^x may be calculated from (13.12)$_{1,2}$. Thus follows the (P, d)-curve of phase equilibrium which in the present case is a descending straight line.

The negative slope of that line indicates that the phase equilibria are unstable. And indeed, it can be shown that

- in fact the free energy on this line possesses a minimum in comparison with other states of the same deformation,
- but that the free enthalpy $f - Pd$ has a maximum on this line in comparison with other states of the same load.

Thus the points on the equilibrium line are similar to saddle points. A state on this line can "slide off" laterally as indicated by the horizontal arrows in Fig. 13.11. It seems plausible to relate the phenomena of internal yield and internal recovery, cf. Fig. 13.9, to this instability.

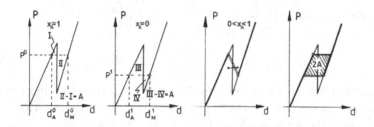

Fig. 13.11. On the construction of the hysteresis loop
Also: The descending line of phase equilibrium

We conclude from these arguments that the pseudoelastic hysteresis is caused by the coherency factor A. If $A = 0$ holds, the condition (13.13) degenerates into the well-known prescription of Maxwell's "equal area rule" for the determination of the load of phase equilibrium. In that case the phase equilibrium is no longer unstable; rather it is indifferent and the hysteresis loop shrinks to a single horizontal line.

13.5 Latent heat

The austenitic $\Longleftrightarrow$ martensitic phase transition during yield and recovery in the pseudoelastic range does not proceed along equilibria. Indeed yield and recovery occur along horizontal lines in the (P, d)-diagram rather than along the descending diagonal equilibrium line of Fig. 13.11. We proceed to calculate the specific heating $\dot{q}$ associated with yield and recovery.

The specific heating ocurs in the 1st and 2nd laws, viz.

$$\frac{du}{dt} - \dot{q} - P\frac{dd}{dt} = 0 \quad \text{and} \quad \frac{ds}{dt} - \frac{\dot{q}}{T} = \sigma \geq 0 , \qquad (13.14)$$

where σ is the non-negative specific entropy source. With $s = -\frac{\partial f}{\partial T}$, and f from (13.11), and f_A, f_M from (13.8) we obtain

$$\frac{ds}{dt} = -\frac{1}{2}k\ln\left(\frac{K_A}{K_M}\right)\frac{dx_A}{dt} , \quad \text{hence by (13.14)}_2 \quad \dot{q} = -\frac{1}{2}k\ln\frac{K_A}{K_M}\frac{dx_A}{dt} - T\sigma .$$

$$(13.15)$$

Therefore it remains to calculate the entropy production σ. Elimination of $\dot{q}$ between $(13.14)_{1,2}$ provides with $f = u - Ts$ — always for an isothermal process —

$$-T\sigma = \frac{df}{dt} - P\frac{dd}{dt} , \quad \text{or by (13.11)}$$

$$-T\sigma = (1 - x_M)\left(\frac{\partial f_A}{\partial d_A} - P\right)\frac{dd_A}{dt} + x_M\left(\frac{\partial f_M}{\partial d_M} - P\right)\frac{dd_M}{dt} -$$

$$- [f_M - f_A - P(d_M - d_A) + A(1 - 2x_M)]\frac{dx_A}{dt} . \qquad (13.16)$$

We assume that force equilibrium prevails, even along the yield and recovery lines, where phase equilibrium does not prevail. By $(13.12)_{1,2}$ and with $\lambda = P$ this means that the underlined terms in (13.16) vanish so that (13.16) may be written in the form

$$-T\sigma = +\left[P(d_M - d_A) - \int_{d_A}^{d_M} Pdd - A(1 - 2x_M)\right]\frac{dx_A}{dt} . \qquad (13.17)$$

On the yield- and recovery lines we have $P = P^0$ and $P = P^1$ respectively, cf. Sect. 13.4 and these load values are given by

$$P(d_M - d_A) - \int_{d_A}^{d_M} Pdd = \pm A$$

respectively. It follows from (13.17) that we have

$$-T\sigma = \begin{cases} +2A(1 - x_A)\dfrac{dx_A}{dt} & \text{yield line} \\ & \text{on the} \\ -2A \quad x_A \dfrac{dx_A}{dt} & \text{recovery line .} \end{cases} \qquad (13.18)$$

Insertion into $(13.15)_2$ determines the heating during yield and recovery

$$\dot{q} = \begin{cases} \left(-\dfrac{1}{2}kT \ln \dfrac{K_A}{K_M} + 2Ax_M \right) & \dfrac{dx_A}{dt} \quad \text{yield} \\[3mm] \left(-\dfrac{1}{2}kT \ln \dfrac{K_A}{K_M} + 2A(1 - x_M) \right) & \dfrac{dx_A}{dt} \quad \text{recovery} . \end{cases} \qquad (13.19)$$

If we run only partly through the yield line beginning at $x_A = 1$ and proceeding to a generic value x_A the integrated heating is given by

$$q^Y(x_M) = \frac{1}{2}kT \ln \left(\frac{K_A}{K_M} \right) x_M - Ax_M^2 . \qquad (13.20)$$

Similarly for partial recovery starting at $x_M = 1$ we obtain

$$q^R(x_M) = \frac{1}{2}kT \ln \left(\frac{K_A}{K_M} \right) (x_M - 1) - A (x_M - 1)^2 . \qquad (13.21)$$

The first terms in (13.20), (13.21) are different in sign so that the heat removed during yield is restored during recovery. That is the well-known reversible heat exchange during an isothermal phase transition, the so-called latent heat. For $A = 0$ that is the only contribution to heating. If, however, $A \neq 0$ holds, the phase transition is accompanied by an irreversible heating which depends quadratically upon the phase fraction.

The total heat production for a complete transition of the hysteresis loop – yield and recovery and all – comes out as

$$q^Y (x_M = 1) + q^R(x_M = 0) = -2A . \qquad (13.22)$$

Not surprisingly it corresponds to the area inside the loop, cf. Fig. 13.11.

13.6 Kinetic theory of shape memory

The basis of the kinetic theory of shape memory is the formulation of rate laws for the phase fractions of the austenitic lattice layers and of the martensitic twins. We continue to consider the potential energy (13.1) with its three minima, cf. Fig. 13.6 and Fig. 13.12. On the left hand side of the left barrier and on the right hand side of the right barrier we have the M_--layers and M_+-layers respectively. Between the barriers we have A-layers. We ask for rates of change of the phase fractions $x_{M\pm}$ and x_A. We assume that the phases are in equilibrium within themselves so that (13.6) holds. Also we suppose that there is equilibrium of forces so that $\beta_\pm$ and β_A are all equal to the load P.

We let the rates be determined by *rate laws* as follows

$$\begin{aligned} \dot{x}_{M_-} &= -p^{-0}x_{M_-} + p^{0-}x_A \\[2mm] \dot{x}_A &= p^{-0}x_{M_-} - p^{0-}x_A - p^{0+}x_A + p^{+0}x_{M_+} \\[2mm] \dot{x}_{M_+} &= \qquad\qquad\qquad\qquad p^{0+}x_A - p^{+0}x_{M_+} \end{aligned} \qquad (13.23)$$

For motivation of this ansatz we discuss the rate $\dot{x}_{M_-}$. It has two parts: a loss and a gain. The loss is due to the layers that jump out of the left potential well; their number is proportional to the number of layers in that well and the factor of proportionality is p^{-0}, the *transition probability* from left to center. The gain is due to the layers that jump from the central potential well into the left one and it is proportional to the fraction of central layers with the factor of proportionality p^{0-}. The other rates in (13.23) are similarly constructed except that $\dot{x}_A$ contains four terms, because the central well can exchange layers with both lateral ones.

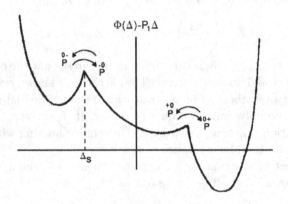

Fig. 13.12. Exchange of lattice layers between potential wells

The transition probabilities are constructed by the principles of theories of activated processes which were first developed in chemistry. In the present case the transition probability p^{-0} thus results as

$$p^{-0} = \sqrt{\frac{kT}{2\pi m}} \; \frac{e^{-\frac{\Phi(\Delta_s)+P\Delta_s}{kT}}}{\displaystyle\sum_{\Delta_{M_-}} e^{-\frac{\Phi(\Delta)-P\Delta}{kT}}} \; e^{-\frac{A}{kT}(1-2x_A)} \qquad (13.24)$$

By (13.6)$_2$ the fraction in this expression determines the probability that a layer of phase M_- has an energy as big as the potential energy at $\Delta = -\Delta_s$. The idea is that a layer cannot surpass the barrier unless it has reached that height at least. Actually the layer needs a little more energy than that: Its energy needs to reach the value $\Phi(-\Delta_s) + P\Delta_s + A(1 - 2x_A)$, because the coherency energy $NAx_A(1 - x_A)$ changes by the amount $A(1 - 2x_A)$ when a layer passes a barrier, cf. Sect. 13.4. This interfacial contribution is taken care of in (13.24) by the exponential term connected with A. The square root in the equation represents the mean velocity with which M_--layers move toward the barrier, where m is an effective mass of the layer.

The sum in the denominator of the fraction in (13.24) may be converted into an integral in the manner described previously in Sect. 13.3 and we obtain

$$p^{-0} = \frac{1}{2\pi Y}\sqrt{\frac{2K_M}{m}}e^{-\frac{P^2/4K_M-PJ}{kT}}e^{-\frac{\Phi(-\Delta_s)+P\Delta_s}{kT}}e^{-\frac{A(1-2x_A)}{kT}}. \tag{13.25}$$

We see in this formula an explicit confirmation of our previous argument about entropic stabilization: When the curvature K_M of the martensitic potential wells becomes bigger, i.e. when the well becomes narrower, the exit probability grows.

The other transition probabilities may be calculated in the same manner as (13.24) or (13.25). We do not list them. Suffice it to say that all transition probabilities are explicit functions of P, T and $x_{M_\pm}$, x_A. Therefore the rate laws (13.23) permit us to solve the following problem:

> Given $P(t)$, $T(t)$ and the initial values $x_{M_\pm}(0)$, $x_A(0)$
> we may calculate $x_{M_\pm}(t)$, $x_A(t)$.

The solution results by stepwise integration in a simple numerical scheme. An analytic solution is impossible because of the ubiquitous non-linearities.

As soon as $x_{M_\pm}(t)$, $x_A(t)$ have been determined we may calculate the deformation $D(t)$. For that purpose we write $D = D_{M_-} + D_A + D_{M_+}$ and calculate D_{M_+} and D_A from (13.7). We recall that β_{M_+} and β_A are all equal to P, cf. (13.12)$_{1,2}$ and obtain

$$D = N\left\{x_{M_-}\left(\frac{P}{2K_M}+J\right)+x_A\frac{P}{2K_A}+x_{M_-}\left(\frac{P}{2K_M}-J\right)\right\}. \tag{13.26}$$

Figure 13.13 shows some solutions $x_{M_\pm}(t)$, $x_A(t)$ and $D(T)$ thus obtained. Obviously for numerical solutions we need to introduce dimensionless variables. As such we have chosen

$$\text{temperature } \Theta = \frac{kT}{\Phi(\Delta_s)}$$

$$\text{load} \qquad p = \frac{J}{\Phi(\Delta_s)}P \tag{13.27}$$

$$\text{deformation } l = \frac{D}{NJ}.$$

Furthermore we need to fix the parameters of the model. We set

$$\frac{A}{\Phi(\Delta_s)} = 0.35, \quad \frac{\Phi_0 + K_A\Delta_s^2}{\Phi_0} = 10.8, \quad \frac{K_M}{K_A} = 100, \quad \frac{1}{\sqrt{2\pi Y}}\sqrt{\frac{K_A}{m}} = 100. \tag{13.28}$$

It must be admitted that these parameters have been chosen by a mix of reasonable expectation and hindsight. They provide us with the "right" type of curves in Fig. 13.13. We proceed to discuss those curves.

In all situations represented by Fig. 13.13 the load is prescribed as a triangular function alternating between tension and compression. The temperature is kept constant: low on top and high at the bottom.

In the upper two tableaux we start with $x_{M_\pm} = \pm\frac{1}{2}$ and observe that x_{M_+} gains at the expense of x_{M_-} as long as the load is a tensile load. When the load becomes compressive, x_{M_-} is formed. The deformation alternates between positive and negative values along with the load. The lower tableaux refer to higher temperatures. Accordingly we start with $x_A(0) = 1$. Under the tensile load of sufficient size the deformation is still alternating between positive and negative values for tensile and compressive loads, but it remains small as long as the austenite prevails.

Particularly instructive are the (P, D)-diagrams of Fig. 13.13 that follow from $P(t)$ and $D(t)$ by the elimination of time. Comparison with the schematic graphs of Fig. 13.2 shows that the model body traces out the quasiplastic hystereses at low temperatures and the pseudoelastic hystereses at high temperature.

13.7 Molecular dynamics

A two-dimensional atomistic model for a shape memory alloy is shown in the two pictures of Fig. 13.14., where two constituents — one with black atoms, the other one with gray atoms — form square and hexagonal lattices, respectively. Each of the N atoms is characterized by a Greek index and all atoms interact with all others by a Lennard-Jones potential dependent on the distance $x_{\beta\alpha} = |\mathbf{x}_\beta - \mathbf{x}_\alpha|$.

$$\Phi_{\alpha\beta}(x_{\beta\alpha}) = 4\varepsilon_{\alpha\beta}\left(\left(\frac{\sigma_{\alpha\beta}}{x_{\beta\alpha}}\right)^{12} - \left(\frac{\sigma_{\alpha\beta}}{x_{\beta\alpha}}\right)^6\right) \qquad (\alpha, \beta = 1, 2, ...N). \quad (13.29)$$

$\varepsilon_{\alpha\beta}$ and $\sigma_{\alpha\beta}$ represent six parameters, because atoms of the two constituents interact differently and so do unlike pairs of atoms. The equations of motion read

$$m_\alpha\ddot{\mathbf{x}}_\alpha = \sum_{\substack{\beta=1 \\ \beta\neq\alpha}}^{N} \frac{\partial\Phi_{\alpha\beta}}{\partial x_{\beta\alpha}} \frac{\mathbf{x}_\beta - \mathbf{x}_\alpha}{x_{\beta\alpha}} \qquad (\alpha, \beta = 1, 2, ...N). \quad (13.30)$$

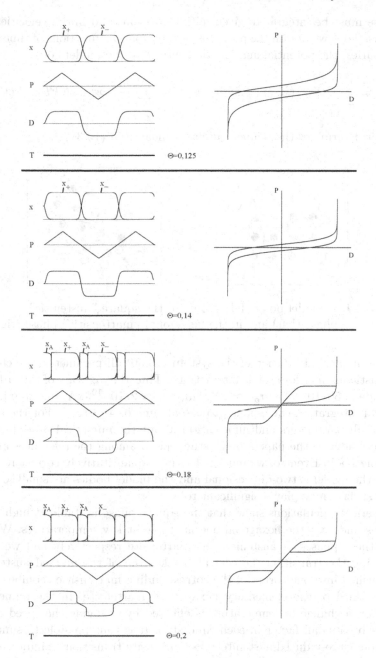

Fig. 13.13. Response of the model body to the application of
a triangular alternating tension and compression
at different constant temperatures

These must be integrated, given initial positions and initial velocities for all atoms. Thus we obtain the positions of the atoms as functions of time and, as corollaries, the potential and kinetic energies of the model body.

$$E_{\text{Pot}} = \frac{1}{2} \sum_{\substack{\beta=1 \\ \beta \neq \alpha}}^{N} \Phi_{\alpha\beta}(x_{\beta\alpha}) \quad \text{and} \quad E_{\text{Kin}} = \sum_{\alpha=1}^{N} \frac{m_\alpha}{2} \dot{x}_\alpha^2 = NkT \; ; \quad (13.31)$$

the latter determines the temperature as indicated in $(13.31)_3$.

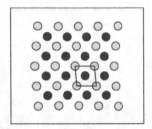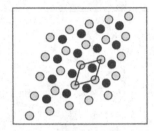

Fig. 13.14. A model body of 41 atoms in the square "austenitic" phase (left) and in the hexagonal "martensitic" phase (right)

In the numerical solution of the system (13.30) all parameters are chosen in a realistic manner for atomic interactions. Thus the m_α's are of the order of magnitude of $50 \cdot 10^{-27}$kg, $\sigma_{\alpha\beta} \approx 10^{-10}$m, $\varepsilon_{\alpha\beta} \approx 2 \cdot 10^{-19}$J. A time step in the numerical integration corresponds to a real time of $4 \cdot 10^{-16}$s. For the exact values of all parameters and information about the numerical procedure we refer the reader to the papers by Kastner, cited among the references at the end of the book. Even so, we must not expect quantitatively correct results, because the model is two-dimensional and the model bodies are small so that the free surface may play a significant role.

Numerical calculations show that the square phase is stable at high temperatures and that the hexagonal one is stable at low temperatures. We interpret these phases as austenite and martensite respectively and we note that during the transition the central black atom of the marked austenitic cell is pushed into one corner of the corresponding martensitic rhombus that covers a third of the elementary hexagon of martensite. In the numerical calculation a change of temperature is effected by changing the speed of all particles by a small factor in each time step. It is thus possible to simulate the temperature-induced austenite $\Longleftrightarrow$ martensite transition as indicated by Fig. 13.15. Inspection shows that the transition austenite $\Longleftrightarrow$ martensite occurs at approximately 400 K while the reverse transition from martensite to austenite occurs at about 700 K.

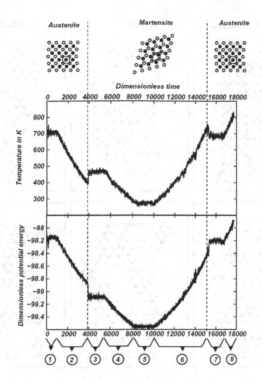

Fig. 13.15. Temperature and potential energy as functions of time during the transitions $A \Longleftrightarrow M$.

The sudden decrease of potential energy and increase in kinetic energy – or temperature – during the transition $A \to M$, which are exhibited by Fig. 13.15, are due to the fact that the potential energy of M is lower than that of A. Therefore, when the body drops to that low level of energy, its atoms accelerate and consequently the temperature grows. The opposite happens during the reverse transition $M \to A$ according to Fig. 13.15.

The transitions $A \Longleftrightarrow M$ themselves may be studied for a body of 230 atoms or a smaller one with 41 atoms on the CD appended to this book, cf. "film 230 ama" or "film 41 ama". We recommend that the reader take a little time and call these movies to the screen and view them.

When the model body is either bigger or when it is fixed at the bottom the temperature-induced transition $A \Longleftrightarrow M$ is accompanied by the formation of martensitic twins, cf. Fig. 13.16. Again, for such cases we have movies on the CD, cf. "film 298 am" or "film 752 am" referring to 298 atoms and 752 atoms respectively.

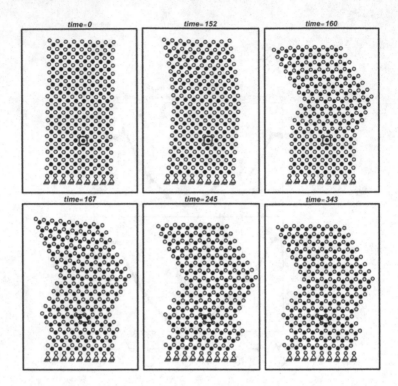

Fig. 13.16. Formation of martensitic twins during a $A \rightarrow M$ transition in a model body of 298 atoms. The atoms at the bottom are fixed in position

13.8 Free energy versus deformation by molecular dynamics under deformation control

We consider the model body of 41 atoms in the situation represented by Fig. 13.17 and prescribe the deformation $D(t) = L(t) - L_0$ of the uppermost atom $\alpha = 41$ as a slowly increasing function of time so that the forces are equilibrated. The phases, however, may not be in equilibrium. The temperature is kept constant by a control algorithm. The figure shows two configurations which can be maintained without a load on atom 41: the rectangular configuration which we call austenite in this model and the rhombic martensite. The load needed to maintain intermediate configurations is generally non-zero and it may be calculated from, cf. (13.30)

$$P_{41} = \sum_{\beta=1}^{40} \frac{\partial \Phi_{41,\beta}}{\partial x_{41,\beta}} \frac{\mathbf{x}_\beta - \mathbf{x}_{41}}{x_{41,\beta}} \ ,$$

since molecular dynamics furnishes the positions of all atoms $\alpha = 1, 2, ... 40$.

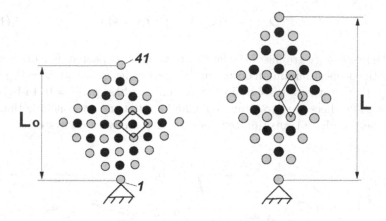

Fig. 13.17. A 41-atom model body in its
austenitic and martensitic phase.

The force $P_{41}(t)$ necessary to maintain a length $L(t)$ depends on the pre-scribed temperature. Thus for small temperatures, where the body "wants" to be martensitic, we expect that a compressive force must be applied in order to prevent the body from spontaneously converting into that phase. For high temperatures on the other hand we expect that a tensile force is needed to pull the body away from the preferred austenitic phase, at least initially. Later, for larger deformations, that force must be decreased and it may even have to become compressive in order to avoid a spontaneous drop into the nearby martensitic configuration. For deformations bigger than the load-free martensitic one all loads must be tensile loads. These expectations are con-firmed by the numerical calculation of P_{41} as shown in Fig. 13.18$_{\text{left}}$, where P_{41} is plotted versus D for different temperatures.

To be sure the graphs of $P_{41}\,(D,T)$ in Fig. 13.18$_{\text{left}}$ are strongly irregular. E.g. the tensile force that maintains configurations close to the austenitic one at high temperature exhibits several kinks. This is due to the fact that the body strives to phase equilibrium, i.e. the microstructure of the model body tends to become inhomogeneous; some of its layers or lattice cells assume the rhombic shape before that is possible for the body as a whole. In that manner the body is able to decrease its elastic energy locally.

The irregularities are less pronounced in the graphs of free energy versus deformation which are shown in Fig. 13.18$_{\text{right}}$. Those graphs are obtained by integration of the $P(D;T)$-graphs

$$F(D,T) - F(0,T) = \int_0^D P(\alpha;T)d\alpha \ . \tag{13.32}$$

The free energy graphs of the figure are much as expected: For low temperatures the deepest minimum occurs in the martensitic phase at $D = D_M$ while at high temperatures it occurs in the austenitic phase at $D = 0$. In-between, at $T = 550K$ there are two minima of equal depths and we conclude that that would be the phase transition temperature if there were no hysteresis.

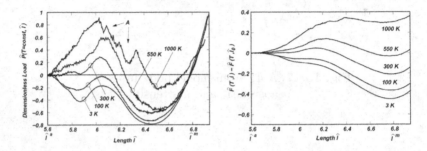

Fig. 13.18. Isotherms of a 41-atom model body under deformation control
Left: Load vs. deformation
Right: Free energy vs. deformation

Since everything is known about all 41 atoms of the model body, we may also plot the potential energy $(13.31)_1$. This is done in Fig. 13.19$_{left}$ and we conclude that the potential energy has its minimum in the martensitic phase *irrespective of temperature*. Therefore energy favours martensite. What stabilizes the austenitic phase at high temperature is again the entropy. Indeed, if we calculate $TS = E_{Pot} - F$ and plot it by superposing the graphs of E_{Pot} and F we obtain the entropy curves exhibited by Fig. 13.19$_{right}$. Inspection shows that the entropy is always largest for $D = 0$, so that entropy favours austenite. And while E_{Pot} depends only weakly on temperature, the entropic contribution to the free energy becomes more and more pronounced the higher the temperature is − essentially because of the explicit factor T in front of S. Therefore we may say that *the austenitic phase is entropically stabilized* at high temperatures.

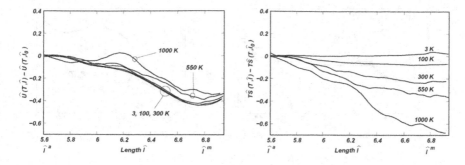

Fig. 13.19. Isotherms of a 41-atom model body
Left: E_{Pot} versus deformation
Right: TS versus deformation

13.9 Molecular dynamics under load control

The 41-atom model body of Fig. 13.14 may also be subject to a slowly increasing prescribed force $P_{41}(t)$. The force increases so slowly that phase equilibrium can be established. In that case the equations of motion read

$$m_\alpha \ddot{x}_\alpha = \sum_{\substack{\beta=1 \\ \beta \neq \alpha}}^{41} \frac{\partial \Phi_{\alpha\beta}}{\partial x_{\beta\alpha}} \frac{x_\beta - x_\alpha}{x_{\beta\alpha}} + P_\alpha(t)\delta_{\alpha 41} \quad .$$

The deformation $D(t)$ of atom 41 results from integration and it is plotted against $P_{41}(t)$ in Fig. 13.20$_{\text{left}}$ for a temperature of $T = 1000\,\text{K}$ which is kept constant. For better orientation we have reproduced the non-monotonic load-deformation curve of the previous deformation controlled experiment of Sect. 13.8. The present experiment is represented by the black curve and we observe a breakthrough at the load which makes the areas between the previous (P, D)-curve and the breakthrough-line equal.

This is what we expect for an equilibrium transition according to the analysis of Sect. 13.4, if the coherency energy vanishes. In other words: *There is no hysteresis.* To be sure, the computer experiment represented in Fig. 13.20 is conducted very slowly: 2000 time steps elapse before a load step. Under those conditions the body has time enough to sample all energetic states available to it and choose the one which is most favourable.

We point out a conspicuous detail of Fig. 13.20$_{\text{left}}$: Before the transition occurs at the dimensionless load $P = 0.26$, there is a "failed attempt" at $P \approx 0.2$. That attempt fails because the right minimum of $F(D, T) - PD$ for that smaller load is still higher than the left minimum, cf. Fig. 13.20$_{\text{right}}$. The accompanying CD has a movie "film 41 am - lk" which exhibits this behaviour for a 41-atom model body.

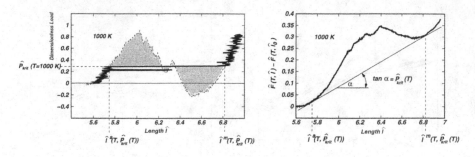

Fig. 13.20. Isotherms of a 41-atom model body under load control
Left: Load vs. deformation
Right: Free energy vs. deformation

13.10 Statistical mechanics and molecular dynamics

We recall the brief description of statistical thermodynamics, cf. Chap.6. And
we consider the body of 221 atoms, shown in Fig. 13.21, as the heat bath
for the marked subbody of $N = 41$ atoms. Thus according to the rules of
statistical thermodynamics the thermodynamic functions U, S and F of the
subbody may be calculated from the partition function

$$P = \sum_{\mathbf{x}_1 \ldots \mathbf{p}_N} e^{-\frac{E(\mathbf{x}_1 \ldots \mathbf{p}_N)}{kT}} \tag{13.33}$$

by use of (6.8). The energy E in (13.33) consists of the sum of kinetic and
potential energies

$$E = \sum_{\alpha=1}^{N} \frac{p_\alpha^2}{2\mu_\alpha} + E_{\mathrm{Pot}}(\mathbf{x}_1 \ldots \mathbf{x}_N) \tag{13.34}$$

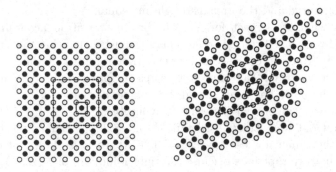

Fig. 13.21 A 41-atomic body in a heat bath
Left: Austenite A, Right: Martensite M.

We anticipate a result that will be derived in Chap. 14 and assume that the number of positions and momenta of atom α beween $\mathbf{x}_\alpha$, $\mathbf{p}_\alpha$ and $\mathbf{x}_\alpha+d\mathbf{x}_\alpha$, $\mathbf{p}_\alpha+d\mathbf{p}_\alpha$ equals $\frac{1}{h^2}d\mathbf{x}_\alpha d\mathbf{p}_\alpha$ where h is the Planck constant. We may thus replace the sums in the partition function by integrals. The integrals over p_α can be calculated and we obtain

$$P_h = \left(\frac{2\pi kT}{h^2}\right)^N \mu_G^{N_G}\,\mu_B^{N_B} \int_{\mathbf{x}_1}..\int_{\mathbf{x}_N} e^{-\frac{E_{Pot}^h(\mathbf{x}_1...\mathbf{x}_N)}{kT}}\,d\mathbf{x}_1...d\mathbf{x}_N, \quad (h = A, M),$$

$$(13.35)$$

where μ_G and μ_B are the masses of the N_G gray and the $N_B = N - N_G$ black atoms respectively. The index h refers to the phases A or M. The integration over the positions poses a problem which we solve approximately as follows:

Inspection of the atomic positions in any of the animations cited in Sects. 13.7 through 13.9 shows that all atoms in either one of the phases fluctuate about fixed positions. They never leave the neighbourhoods of those positions. Thus Fig. 13.22 shows 5000 consecutive positions of the atoms of a single austenitic cell (left) and a single martensitic cell (right). We see that the clouds of dots are roughly circular in both cases and that the austenitic dots are bigger than the martensitic ones. In fact, if we plot the probability of the distance of dots from the central point, we obtain graphs like those of Fig. 13.23 which may be approximated by Gaussians – a wide and flat one for the austenitic phase A and a narrow and steep one for the martensite M.

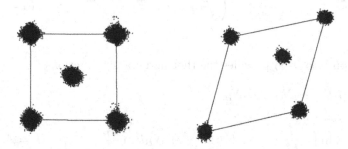

Fig. 13.22 Clouds of 5000 dots marking consecutive positions of the atoms of one lattice cell.
Left: Austenite A. Right: Martensite M.

Fig. 13.23 is measured for a black atom within a cell of gray ones. Qualitatively similar curves result for gray atoms in black cells. This suggests that an inner atom in a lattice cell finds itself in a paraboloidal potential well with curvature λ and minimum ε at $\bar{x}$. We set

$$E_i^h(\mathbf{x}) = \frac{\lambda_i^h}{2}(\mathbf{x} - \bar{\mathbf{x}}_i^h)^2 + \varepsilon_i^h \qquad \begin{matrix} (h = A, M) \\ (i = G, B) \end{matrix}. \qquad (13.36)$$

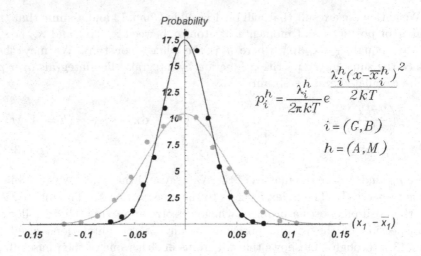

Fig. 13.23 Probability of distance of a fluctuating atom from
the mean position
Flat curve: Austenite. Steep curve: Martensite.
Probability distribution, cf. text.

We may then calculate the partition function of the phases as[3]

$$P_h = \left(\frac{2\pi kT}{h^2}\right)^{2N} \left(\frac{\mu_G}{\lambda_G^h}\right)^{N_G} \left(\frac{\mu_B}{\lambda_B^h}\right)^{N_B} e^{-\frac{\varepsilon_G^h N_G + \varepsilon_B^h N_B}{kT}} \qquad (h = A, M).$$

(13.37)

Insertion into (6.8) provides the thermodynamic functions

$$U_h = 2NkT + \varepsilon_G^h N_G + \varepsilon_B^h N_B,$$

$$S_h = 2Nk \ln\left(\frac{2\pi kT}{h^2}e\right) + N_G k \ln\frac{\mu_G}{\lambda_G^h} + N_B k \ln\frac{\mu_B}{\lambda_B^h}, \qquad (h = A, M)$$

$$F_h = -2NkT \ln\frac{2\pi kT}{h^2} - N_G kT \ln\frac{\mu_G}{\lambda_G^h} - N_B kT \ln\frac{\mu_B}{\lambda_B^h} - \left(\varepsilon_G^h N_G + \varepsilon_B^h N_B\right).$$

(13.38)

The curvatures λ_i^h are known from the numerical calculations and we have
listed them in Table 13.1.

---

[3] In this argument we treat the gray atoms like the black ones, although they lie
at the border of the subbody. It is clear that this introduces a mistake which,
however, we ignore.

Table 13.1   Pertinent data from molecular dynamics

	λ_G $10^3\,\mathrm{J/m^2}$	λ_B $10^3\,\mathrm{J/m^2}$
austenite	2.28	3.70
martensite	7.72	10.62

We calculate the differences

$$U_A - U_M = N_G(\varepsilon_G^A - \varepsilon_G^M) + N_B(\varepsilon_B^A - \varepsilon_B^M)$$

$$(13.39)$$

$$T(S_A - S_M) = N_G kT \ln \frac{\lambda_G^M}{\lambda_G^A} + N_B kT \ln \frac{\lambda_B^M}{\lambda_B^A} = T \cdot 6.54 \cdot 10^{-22} \frac{\mathrm{J}}{\mathrm{K}}$$

and determine the expression with ε's in $(13.39)_1$ from the requirement that $F_A - F_M$ falls below zero at $T = 550$ K, the temperature predicted by Fig. 13.18 as the phase transition temperature. Thus we obtain

$$U_A - U_M = 3.60 \cdot 10^{-19} \text{ J}. \qquad (13.40)$$

We conclude from (13.40) and $(13.39)_2$ that
- the martensite is energetically preferred, because it has the lower energy
- the austenite is entropically preferred, because it has the higher entropy.

Since increasing temperature emphasizes the entropic effect, the austenite is the high temperature phase which confirms all our previous findings.

We confirm in particular that the austenite is entropically stabilized at high temperatures. And once again this is basically due to the fact that the atoms in the austenitic cell have more room than in the martensitic one, cf. Fig. 13.22, and entropy increases with increasing volume, or − in the present case − with increasing area.

The third law of thermodynamics.
Capitulation of entropy

It is obvious that as temperature decreases, the entropy S plays a smaller and smaller role in minimizing the available free energy $\mathfrak{A}$. This is so, because entropy enters into $\mathfrak{A}$ as $-TS$. However, the 3rd law implies more: The entropy $S(T,p)$ itself tends to zero with T, irrespective of p. This is so, because even tiny potential barriers between different states become unsurmountable for the body when the thermal motion becomes ever weaker.

14.1 Entropy constant and Planck constant

According to (3.38) the molar entropy constant of a monatomic gas α namely

$$\tilde{s}_R^{\alpha} = \tilde{R}\ln\left(\frac{(kT_R)}{p_R}\, e^{5/2}\frac{XY}{\mu_{\alpha}^3}\sqrt{2\pi\mu_{\alpha}}^{\,3}\right) \tag{14.1}$$

is indeterminate since it contains the factor XY. That factor, or rather its inverse $1/XY$, quantizes the $(\mathbf{x}, \mathbf{c})$-space by giving the size of the smallest element $d\mathbf{x}d\mathbf{c}$, i.e. the element that can accommodate *one* occupiable point. Physicists prefer to introduce the factor μ_{α}^3/XY which quantizes the *phase space* — spanned by coordinates $\mathbf{x}$ and momenta $\mathbf{p} = \mu_{\alpha}\mathbf{c}$ — because they have good reasons to believe that μ_{α}^3/XY is universal, i.e. independent of the nature of the gas, or of the mass μ_{α} of its molecules. They denote this factor by h^3, where h is the Planck constant. Thus (14.1) reads

$$\tilde{s}_R^{\alpha} = \tilde{R}\ln\left(\frac{(kT_R)^{5/2}}{p_R}\, e^{5/2}\frac{1}{h^3}\sqrt{2\pi\mu_{\alpha}}^{\,3}\right). \tag{14.2}$$

We shall attempt to calculate the value of h from thermodynamic data alone, without reference to quantum mechanics.

For that purpose we consider the chemical reaction $H \rightarrow \frac{1}{2}H_2$. Measurements of the heat of reaction have provided the value $\Delta\tilde{h}_R = -215.9\frac{kJ}{mol}$. And — by (12.11) — the entropy of reaction is related to the measured and

Table 14.1 Measured partial pressure relations in equilibrium
as function of temperature for the reaction $H \rightarrow \frac{1}{2}H_2$.

T	$\dfrac{\sqrt{p_R p_{H_2}}}{p_H}$	$\ln \dfrac{\sqrt{p_R p_{H_2}}}{p_H}$
1500 K	$5.75 \cdot 10^4$	10.96
2200 K	$1.74 \cdot 10^2$	5.16
3000 K	6.54	1.88

tabulated partial pressures in chemical equilibrium. A brief excerpt of such a
table – for the reaction $H \rightarrow \frac{1}{2}H_2$ – is given in Table 14.1.

The data of any row in Table 14.1 may be inserted into (12.11) and thus
we obtain

$$\Delta \tilde{s}_R = -\tilde{s}_R^H + \frac{1}{2}\tilde{s}_R^{H_2} = -48.16 \frac{J}{\text{mol K}}. \tag{14.3}$$

We introduce the entropy constants (14.2), appropriate to H and H_2, into
this relation and solve for h^3

$$h^3 = \frac{1}{2^{3/2}} \frac{(kT_R)^{5/2}}{p_R} e^{5/2} \sqrt{2\pi \mu_H}^3 \, e^{\frac{2\Delta \tilde{s}_R}{R}} = 467.3 \cdot 10^{-102} \, (\text{Js})^3$$

$$h = 7.76 \cdot 10^{-34} \text{Js} . \tag{14.4}$$

Hence follow the entropy constants by insertion into (14.2)

$$\tilde{s}_R^H = 105.0 \frac{J}{\text{mol K}} \quad \text{and} \quad \tilde{s}_R^{H_2} = 113.68 \frac{J}{\text{mol K}} . \tag{14.5}$$

All values in (14.4), (14.5) are of the right order of magnitude but neither
one is quantitatively good; in particular h is known to quantum physicists to
have the value $6.6252 \cdot 10^{-34}$ Js. And yet, we find it remarkable that from
purely non-quantum data we can obtain a value for h in (14.4) which is at
least close to the correct value. This is as far as thermodynamics can carry us.
For anything more accurate we need quantum mechanics, or rather quantum
thermodynamics.

The reason for the discrepancy lies in the fact that the atom H and, in particular, the molecule H_2 can store energy in "modes" other than the kinetic energy of
translation. Therefore the H-atom and the H_2-molecule have entropies of their own.
This is even more true for more complex atoms and molecules so that the above
value of h, – calculated from $H \rightarrow \frac{1}{2}H_2$, –, bad as it is quantitatively, is still the
best one to be had from thermodynamic data.

Having discussed the reason for the discrepancy we now proceed to use the correct value for h and calculate the corresponding entropy constant for atomic hydrogen from (14.2). We obtain

$$h = 6.6252 \cdot 10^{-34} \, Js \quad \text{and} \quad \tilde{s}_R^H = 108.9 \frac{kJ}{mol \, K} \, . \tag{14.6}$$

This value of the entropy constant $\tilde{s}_R^H$ is still not satisfactory, because it neglects the entropy of the atom at rest which, according to (6.12), has the value $\check{R} \ln 2 = 5.76 \, \frac{J}{mol \, K}$. That value has to be added to (14.6) and thus we find the correct value

$$\tilde{s}_R^H = 114.66 \frac{kJ}{mol \, K} \, , \text{ hence by (14.3)} \quad \tilde{s}_R^{H_2} = 133.0 \frac{kJ}{mol \, K} \, . \tag{14.7}$$

Note that in this manner we have determined the entropy constants *themselves*, — not only their combination represented by the entropy of reaction $\Delta \tilde{s}_R = \sum_\alpha \gamma_\alpha \tilde{s}_\alpha^R$.

There is a similar situation with respect to the energy constants $\tilde{h}_R^\alpha$ which we can know *separately* — in principle — rather than only their combination $\Delta \tilde{h}_R = \sum_\alpha \gamma_\alpha \tilde{h}_R^\alpha$ in the heat of reaction. The energy constant is determined by the mean mass m_R^α of the particles in the reference state via the Einstein formula $\tilde{h}_R^\alpha = m_R^\alpha c^2$. To be sure this significant formula is impractical for the calculation of $\tilde{h}_R^\alpha$, because it is difficult to measure the mean mass m_R^α accurately. Yet the fact remains that, by coincidence, the first decade of the 20th century furnished both: the theoretical possibility of the determination of the absolute values of energy and entropy.

14.2 Entropy at absolute zero

Once the entropy constants $\tilde{s}_R^H$ or $\tilde{s}_R^{H_2}$ are known by (14.7), the entropy itself is known for any pair (T, p), say (T_0, p_0), for which H and H_2 are ideal gases. The appropriate formula for H_2 reads

$$\tilde{s}^{H_2}(T_0, p_0) = \check{R} \left[\frac{7}{2} \ln \frac{T_0}{T_R} - \ln \frac{p_0}{p_R} + \tilde{s}^{H_2}(T_R, p_R) \right] . \tag{14.8}$$

Starting from that value we may calculate the entropy for lower values of T — and for the same value of p, viz. p_0 — by integrating over $\frac{\tilde{c}_p(T, p_0)}{T}$ with measured values of $\tilde{c}_p(T, p_0)$

$$\tilde{s}^{H_2}(T, p_0) = \tilde{s}^{H_2}(T_0, p_0) + \int_{T_0}^{T} \frac{\tilde{c}_p(\tau, p_0)}{\tau} d\tau$$

and this holds as long as our isobaric path of integration does not cross a line of phase transition. If, however, at $T = T_1$ we experience condensation and at $T = T_2$ solidification, we obtain for $T < T_2 < T_1 < T_0$

$$\tilde{s}^{H_2}\left(T, p_0\right) = \tilde{s}^{H_2}\left(T_0, p_0\right) + \int_{T_0}^{T} \frac{\tilde{c}_p(\tau, \, p_0)}{\tau} d\tau - \sum_{i=1}^{2} \frac{r(T_i)}{T_i}, \qquad (14.9)$$

where $r(T_i) > 0$ are the latent heats of the phase transitions. In this manner we can obtain $\tilde{s}^{H_2}\left(T, p_0\right)$ for low and lowest temperatures from measurements of latent and specific heats.

If this experimental program is carried out, it turns out that

$$\boxed{\tilde{s}\left(T, p_0\right) \underset{T \to 0}{\to} 0 \quad \forall \; p_0 \qquad \begin{array}{l} 3^{\text{rd}} \text{ law of} \\ \text{thermodynamics.} \end{array}}$$

$$(14.10)$$

This holds not only for H_2 but for all materials that adopt a crystalline solid phase. [To be sure in some materials we pass through more than one crystalline phase as the temperature is lowered, but that only means that there are more than just two terms in the sum of (14.9).]

14.3 The capitulation of entropy

In all available free energies we have seen that the entropy enters in the combination $U - TS$. Therefore we expect that the entropy matters less when the temperature decreases and we have demonstrated in several instructive cases — starting with the planetary atmosphere in Chap. 8 — that this is indeed the case.

But the third law of thermodynamics implies more than that. Indeed, the entropy *itself* — not only its factor T in the free energy — becomes smaller and smaller and tends to zero with $T \to 0$. We proceed to explain that phenomenon in a plausible manner. The key to the understanding lies in the knowledge that the particles of a body are kicked around — by the thermal motion — among the energy states *that are available to them*.

When the temperature becomes smaller, the thermal motion becomes less intensive and terms with high energies become unavailable and cease to contribute to the partition function (6.8), and therefore to entropy. At really small temperatures even the smallest energetic barriers cannot be overcome by the thermal motion, so that the particles settle down in their energetic minima. Thus the atoms of the crystalline lattice into which hydrogen solidifies at low temperature have only *one* position available to them: The point at the bottom of the potential well, where the atom has the lowest energy.[1] The partition function is therefore equal to 1; thus $\ln P$ equals zero and so does the entropy, cf. (6.8).

---

[1] We forego a discussion of the indeterminacy principle here by which a particle cannot lie in a point without having infinite momentum.

We may say then that the *energy* determines the position of atoms and therefore the shape of the body. *Entropy* has capitulated, it is zero and loses all influence on the shape of the body.

Nor do the electrons within the atoms have two zero energy states to choose from, with spin up or down respectively, as they did in a free atom. Indeed, in the low temperature *crystal* each atomic spin interacts with the spins of the neighbouring atoms and it takes much energy to switch the spin from one direction to the other one.

Low temperature physics is a fascinating field of research at the present time and the third law is a firmly established axiom of that field. We are only scratching the surface of that subject when we talk about the electrons in the hydrogen crystal. Indeed the spins of the nuclei must be taken into account and their interactions with all spins of the neighbouring atoms.

So firmly do low-temperature-physicists believe in the third law that they predict that two copper isotopes – Cu^{63} and Cu^{65} – should unmix at very low temperature so as to shed their entropy of mixing. Of course, the mixture is solid and unmixing should take a long time and would be hard to observe.

What *is* observed is the unmixing of He^3 and He^4, both liquids at temperatures below 0.8 K, so that He^3 floats above He^4. Thus entropy is decreased by getting rid of the entropy of mixing. We must assume that there is an energetic benefit for the mixture in the unmixing process in order to offset the decrease of the entropy of mixing. That benefit seems to lie in the energetic interaction of the nuclear spins of He^3.

15

The zeroth law of thermodynamics — kinetic and thermodynamic temperatures

The zeroth law of thermodynamics defines the *thermodynamic temperature* T as *the* quantity that is continuous at an interface between two bodies. This definition is essentially the same as saying that T is the factor of proportionality between heat flux and entropy flux.

On the other hand we often define the temperature as a measure for the mean kinetic energy of the atoms of a body. We might call that quantity the *kinetic temperature*. In equilibrium both definitions are equivalent, but there is a difference in non-equilibrium. In this chapter we demonstrate this observation.

We also show that, according to the kinetic theory of gases, there is a tensorial relation between heat flux and entropy flux. Indeed, for a strong degree of rarefaction of the gas, the entropy flux is not related to the heat flux alone. Thermodynamics has not, however, progressed far enough to evaluate such cases.

15.1 Continuity of temperature and the entropy flux

The zeroth law of thermodynamics is a statement on temperature. It says that the temperature is continuous at the interface between two bodies

$$[T] = 0 \quad \begin{array}{l} \text{zeroth law of} \\ \text{thermodynamics.} \end{array}$$

(15.1)

Actually this is the *defining property* of temperature. Square brackets $[a]$ represent the difference of a on the two sides of the interface; for temperature that difference is zero.

Some people will insist that the continuity of temperature is valid only in equilibrium, but everybody uses it for the calculation of the temperature field in a layered wall, like a thermopane window, irrespective of the size of the heat flux passing the wall or the window. Thus it seems that the continuity of T is credited — and creditable — even outside equilibrium.

Actually there is a neat way to "derive" the zeroth law, or rather to find an equivalent definition that — to some people — may clarify the status of temperature: Let us consider an interface between two bodies, and let the interface neither accumulate entropy nor produce entropy. On such an interface, with the unit normal **n**, the normal component of the entropy flux vector **Φ** is continuous, i.e. the influx of entropy on one side equals the outflux on the other side. We have

$$[\Phi_i] \, n_i = 0 \, . \tag{15.2}$$

Now, let us assume that the entropy flux Φ_i and the heat flux, or flux of internal energy q_i are related by

$$\Phi_i = \frac{q_i}{T} \tag{15.3}$$

as suggested by the 2nd law. If this is so, we obtain from (15.2)

$$\left[\frac{q_i}{T}\right] \, n_i = 0, \quad \text{hence} \quad [T] = 0 \, , \tag{15.4}$$

provided that the interface does not accumulate internal energy either. Thus the temperature is indeed continuous irrespective of the value of the heat flux.

It is usually considered that the assumptions on the absence of accumulation and production of entropy and internal energy are well-satisfied, e.g. for the wall of a thermometer, and therefore the relation $\Phi_i = \frac{q_i}{T}$ is equivalent to the zeroth law. In a manner of speaking we can thus say that $\Phi_i = \frac{q_i}{T}$ is an alternative *definition* of temperature T.

We may ask of course at this stage how much value a definition does have that involves the entropy flux, which is surely a most esoteric quantity?

15.2 Temperature as a measure for the mean kinetic energy

And yet there is a case for which we do know the entropy flux, namely the case of monatomic ideal gases according to the kinetic theory of gases, cf. Chap. 3. We ask whether $\Phi_i = \frac{q_i}{T}$ is confirmed by the kinetic theory and the immediate answer is: No! Indeed, equation (3.25) shows that there is an additional term due to the non-equilibrium part φf_E of the distribution function.

However, upon reflection there is a more profound answer, perhaps, to the question. That new answer starts from the observation that the temperature in the kinetic theory is independently defined as a measure for the mean kinetic energy of the atoms, cf. (3.10) and it is not *a priori* clear whether that temperature satisfies the zeroth law. To be sure, in equilibrium it does; that is a fact amply proven by statistical mechanics. But what about non-equilibrium?

This is the question to be answered in this chapter and, in order to highlight it, we denote the kinetic temperature by ϑ − rather than T. Thus we rewrite (3.25) in the form

$$\Phi_i = \frac{q_i}{\vartheta} - k \int C_i \ln(1 + \varphi) f \mathrm{dc}. \tag{15.5}$$

Before we can do anything with this formula we have to determine φf_E, the non-equilibrium part of the distribution function.

15.3 Grad's 13-moment theory

13-moment distribution function

Grad has proposed to write non-equilibrium distribution functions as expansions of f/f_E in terms of Hermite-polynomials. The simplest realistic one among these is the 13-moment distribution function

$$f = \left(1 + \frac{1}{\left(\frac{k}{\mu}\vartheta\right)^2} \frac{1}{2\rho} p_{\langle ij\rangle} C_i C_j + \frac{1}{\left(\frac{k}{\mu}\vartheta\right)^3} \frac{1}{5\rho} q_i \left(C^2 - 5\frac{k}{\mu}\vartheta \right) C_i \right) f_E, \quad (15.6)$$

where q_i is the heat flux and $p_{\langle ij\rangle}$ is the deviatoric pressure tensor, i.e. the traceless part of p_{ij}, cf. (3.8). f_E is the Maxwell distribution (3.17) which we now write in terms of the kinetic temperature ϑ rather than T, as explained.

13-moment entropy flux

Comparison of (15.6) with (3.24) identifies φ and, when we introduce that value of φ into (3.25), and set $\ln(1 + \varphi) \approx \varphi$, we obtain

$$\Phi_k = \frac{q_k}{\vartheta} - \frac{2}{5} \frac{p_{\langle ki\rangle}}{\rho \frac{k}{\mu}\vartheta} \frac{q_i}{\vartheta} . \quad (15.7)$$

Therefore the entropy flux is not proportional to the heat flux. It is then not possible, in general, to define a thermodynamic temperature T and therefore (15.3) cannot be used to relate T and ϑ. A little later we shall consider a special, highly symmetric case where this may be done, but it cannot be done in general.

It is true that the leading term in (15.7) is q_i/ϑ but that term is not the only one. It is also true that, according to (15.7), there is still a linear relation between Φ and q, but the relation is a tensorial one. Researchers in thermodynamics have not progressed yet to an interpretation of what that means for thermodynamics and, in particular, for the 2nd law and the concept of temperature.

There is an analogy to the present situation and that concerns diffusion through a permeable wall or a phase boundary. In that case the entropy flux has a term proportional to the diffusion flux and in simple cases the chemical potential is the factor of proportionality. [The chemical potential has much in common with temperature; in particular, it is continuous at a permeable wall just as temperature is continuous at a heat-conducting wall.] But, when different non-isotropic stresses are exerted on the adjacent bodies, there is a tensorial relation between the entropy flux and the diffusion flux and we may then define a *chemical potential tensor*. Actually the metallurgists concerned with such cases speak of the *Eshelby tensor*, which is continuous only in its normal components.

One last remark: If we refine the characterization of non-equilibrium, and go beyond the 13-moment distribution function to more-moment Grad distributions, the entropy flux will contain terms that are not dependent on the heat flux at all, but rather on higher moments of the distribution function and combination of them. Such situations, sofar, have been beyond the scope of any thermodynamic theory.

We now turn back to (15.7) and exploit that relation for a case for which it *can* be exploited. This involves the calculation of $p_{\langle ki \rangle}$ and for that we rely on Grad's 13-moment field equation.

13-moment field equations

Apart from the 5 conservation laws (3.7), (3.8) for mass, momentum and energy the Grad 13-moment theory makes use of 8 additional field equations. Those result from the generic equation of transfer (3.5) for $\psi = \mu \left(C_i C_j - \frac{1}{3} C^2 \delta_{ij} \right)$ and $\psi = \frac{\mu}{2} C^2 C_i$. We are not writing these equations in full generality but only for stationary fields ρ, v_i, ϑ, $p_{\langle ij \rangle}$ and q_i and for a gas at rest, i.e. with $v_i = 0$. In that case the mass balance is identically satisfied and the conservation laws for momentum and energy simply state that the divergences of the pressure tensor p_{ij} and the heat flux q_i vanish.

$$\frac{\partial p_{ij}}{\partial x_j} = 0 \quad \text{and} \quad \frac{\partial q_i}{\partial x_i} = 0 . \tag{15.8}$$

The additional 8 equations of transfer we simplify by replacing the complicated production term of the Boltzmann equation by the BGK ansatz $\frac{1}{\tau}(f_E - f)$, where τ is a constant time of the order of the mean time of free flight of the atoms. Under these simplifiying assumptions the additional equations of transfer read

$$\frac{\partial p_{\langle ij \rangle l}}{\partial x_l} = -\frac{1}{\tau} p_{\langle ij \rangle} \quad \text{and} \quad \frac{\partial p_{ilkk}}{\partial x_l} = -\frac{2}{\tau} q_i , \tag{15.9}$$

where p_{ijk} and p_{ijkl} are moments of rank 3 and 4 of the distribution function, viz.

$$p_{ijk} = \int \mu c_i c_j c_k f d\mathbf{c} \quad \text{and} \quad p_{ijkl} = \int \mu c_i c_j c_k c_l f d\mathbf{c} . \tag{15.10}$$

$p_{\langle ij \rangle k}$ is the part of (15.9) that is traceless in i, j and p_{ijkk} is the trace of p_{ijkl} in the indices k, l. Both are calculated in the Grad 13-moment theory by use of the distribution function (15.6) so that we have

$$p_{\langle ij \rangle l} = \frac{2}{5} \left(q_i \delta_{jl} + q_j \delta_{il} - \frac{2}{3} q_l \delta_{ij} \right) \quad \text{and} \quad p_{ilkk} = 5\rho \left(\frac{k}{\mu} \vartheta \right)^2 \left(\delta_{il} + \frac{7}{5} \frac{p_{\langle il \rangle}}{\rho \frac{k}{\mu} \vartheta} \right) . \tag{15.11}$$

Thus Grad's 13-moment field equations in the present case represent a closed set of non-linear differential equations for the fields $\rho(\mathbf{x})$, $\vartheta(\mathbf{x})$, $p_{\langle ij \rangle}(\mathbf{x})$ and $q_i(\mathbf{x})$. We proceed to find a solution.

15.4 Heat conduction in a gas at rest between concentric cylinders

We write the Grad 13-moment system for a stationary process in a gas at rest in a form appropriate for curvilinear coordinates as

$$p^{il}{}_{;l} = 0, \quad q^l{}_{;l} = 0, \quad p^{\langle ij \rangle l}{}_{;l} = -\frac{1}{\tau}p^{\langle ij \rangle}, \quad p^{ilk}{}_{k;l} = -\frac{2}{\tau}q^i \tag{15.12}$$

The semicolon represents covariant derivatives and upper and lower indices indicate contra- and covariant components respectively. The closure relations (15.11) read in curvilinear coordinates

$$p^{\langle ij \rangle l} = \frac{2}{5}\left(q^i g^{jl} + q^j g^{il} - \frac{2}{3}q^l g^{ij}\right) \quad \text{and} \quad p^{ilk}{}_k = 5\rho\left(\frac{k}{\mu}\vartheta\right)^2\left(g^{il} + \frac{7}{5}\frac{p^{\langle il \rangle}}{\rho\frac{k}{\mu}\vartheta}\right), \tag{15.13}$$

where g^{il} is the metric tensor.

We are going to solve these equations for a gas between concentric cylinders. Therefore it is appropriate to use cylindrical coordinates (r, ϑ, z) and the corresponding metric tensor and Christoffel symbols

$$g^{ik} = \begin{pmatrix} 1 & 0 & 0 \\ 0 & \frac{1}{r^2} & 0 \\ 0 & 0 & 1 \end{pmatrix}, \quad \Gamma^r_{\vartheta\vartheta} = -r, \quad \Gamma^\vartheta_{\vartheta r} = \Gamma^\vartheta_{r\vartheta} = \frac{1}{r}, \quad \Gamma^m_{kn} = 0 \quad \text{else.} \tag{15.14}$$

We assume that all fields depend only on r and that q^r is the only non-vanishing component of the heat flux.

The system (15.12) then assumes the simple form

$$\frac{\partial p}{\partial r} = 0, \qquad \text{where} \quad p = \rho\frac{k}{\mu}\vartheta \ \sim \ \text{pressure}$$

$$\frac{\partial q^r}{\partial r} + \frac{1}{r}q^r = 0,$$

$$p^{\langle rr \rangle} = -\tau\frac{4}{5}\frac{\partial q^r}{\partial r}, \qquad p^{\langle \vartheta\vartheta \rangle} = -\tau\frac{4}{5}\frac{1}{r^3}q^r, \tag{15.15}$$

$$q^r = -\frac{5}{2}p \ \tau\frac{\partial \frac{k}{\mu}\vartheta}{\partial r} - \frac{7}{2}p^{\langle rr \rangle} \ \tau\frac{\partial \frac{k}{\mu}\vartheta}{\partial r}.$$

All other quantities vanish and all other equations are identically satisfied. Hence follows the general solution

$$p = p_0 = \text{const.}$$

$$q^r = \frac{c_1}{r}$$

$$p^{\langle rr \rangle} = \tau\frac{4}{5}\frac{c_1}{r^2}, \qquad p^{\langle \vartheta\vartheta \rangle} = -\tau\frac{4}{5}\frac{c_1}{r^4} \tag{15.16}$$

$$q^r = -\frac{5}{2}p\tau\left(1 + \frac{7}{5}\frac{p^{\langle rr \rangle}}{p}\right)\frac{\partial \frac{k}{\mu}\vartheta}{\partial r}.$$

The leading term in the least equation represents the Fourier law by which heat flux and temperature gradient are proportional with $\kappa = \frac{5}{2}p\tau\frac{k}{\mu}$ as factor of proportionality. That law is modified by Grad's 13-moment theory and the modification is inherent in the $p^{\langle rr \rangle}$-term.

The temperature field may easily be determined by integration of $(15.16)_5$ with $(15.16)_2$. We obtain

$$\frac{k}{\mu}\vartheta(r) - \frac{k}{\mu}\vartheta(r_R) = -\frac{c_1}{5p_0\tau}\ln\frac{r^2 + \dfrac{28}{25}\dfrac{\tau c_1}{p_0}}{r_R^2 + \dfrac{28}{25}\dfrac{\tau c_1}{p_0}} . \tag{15.17}$$

The coefficient c_1 is a constant of integration and $\vartheta(r_R)$ is the kinetic temperature at some reference radius r_R. Both must be determined from boundary values. We proceed to do that.

15.5 Thermodynamic and kinetic temperatures

We recall the relation (15.7) for the entropy flux. For heat conduction between concentric cylinders – the problem just solved – that relation reads

$$\Phi^r = \frac{1}{\vartheta}\left(1 - \frac{2}{5}\frac{p^{\langle rr \rangle}}{p}\right)q^r . \tag{15.18}$$

Thus by the zeroth law in the form (15.3) we find a relation between the thermodynamic temperature T and the kinetic temperature, viz.

$$T = \frac{\vartheta}{1 - \dfrac{2}{5}\dfrac{p^{\langle rr \rangle}}{p}} . \tag{15.19}$$

Therefore the kinetic temperature and the thermodynamic temperature agree in equilibrium as expected. But they differ in non-equilibrium. We proceed to calculate the size of the difference for the concentric cylinders treated heretofore, cf. Sect. 15.4.

Since it is the thermodynamic temperature that is continuous at an interface, we can only prescribe $T_{i,o}$ at the inner and outer radii $r_{i,o}$ – and *not* $\vartheta_{i,o}$. The kinetic temperatures $\vartheta_{i,o}$ follow from (15.19) and $(15.16)_{3,5}$

$$\vartheta_{i,o} = T_{i,o}\left(1 - \frac{8}{25}\frac{\tau}{p_0}\frac{c_1}{r_{i,o}^2}\right) \tag{15.20}$$

and therefore the constant c_1 and hence $\vartheta(r_R) = \vartheta_i$ in (15.17) are determined from the equation

$$\frac{k}{\mu}T_i\left(1 - \frac{8}{25}\frac{\tau c_1}{p_0 r_i^2}\right) - \frac{k}{\mu}T_o\left(1 - \frac{8}{25}\frac{\tau c_1}{p_0 r_o^2}\right) = \frac{c_1}{5p_0\tau}\ln\frac{r_o^2 + \frac{28}{25}\frac{\tau c_1}{p_0}}{r_i^2 + \frac{28}{25}\frac{\tau c_1}{p_0}} \quad (15.21)$$

For neon with $p_0 = 10^2\frac{N}{m^2}$ and $\tau = 2.56 \cdot 10^{-7}$s we consider boundary values

$$T_i = 350 \text{ K at } r_i = 2 \cdot 10^{-4} \quad \text{and} \quad T_o = 300 \text{ K at } r_o = 10^{-2} \text{ m.}$$

Hence follows $c_1 = 0.32\frac{N}{s}$ and, by (15.20), $\vartheta_i = 347.7$ K.

With these values we have plotted $T(r)$ and $\vartheta(r)$ in Fig.15.1. Inspection shows that the two temperatures T_i and ϑ_i differ by 2.3 K at the inner radius − not a negligible amount! Thus we conclude that the thermodynamic temperature, defined by the zeroth law, and the kinetic temperature, defined as a measure of kinetic energy, differ.

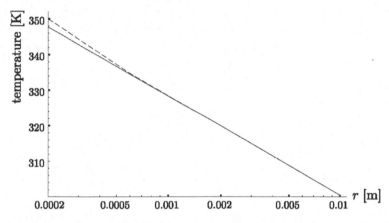

Fig. 15.1 Kinetic (solid line) and thermodynamic (dashed line) temperature of a rarefied neon gas at rest between two co-axial cylindrical shells as a function of the radius.

The deviation of ϑ and T is only noticeable near the inner cylinder, where both fields are steep. This must not surprise us, because the whole effect is due to the terms brought into the theory by the 13-moment theory which differs from the usual Navier-Stokes-Fourier theory only for rarefied gases and steep gradients. Note that in the Grad theory the deviatoric pressure does not vanish in a gas at rest, while in the Navier-Stokes theory it does vanish as a matter of course.

Gibbs paradox and degenerate gases

The easiest way to take care of a paradox is to firmly close the eyes and claim that it does not exist — or does not exist anymore. This usually happens to the Gibbs paradox. To be sure the situation is complex. To begin with, there are two Gibbs paradoxes, one in thermodynamics of mixtures and the other one in statistical mechanics. The latter one is due to an overinterpretation of Boltzmann's formula for the entropy, cf. Sects. 3.6 and 3.7.

Here we review the situation and show how the Gibbs paradox of statistical mechanics can be resolved. The argument provides the opportunity to speak about degenerate gases and the reason for degeneracy.

16.1 Gibbs paradox for the entropy of mixing

We consider ν pure ideas gases α, all at temperature T, pressure p and in the individual volumes V_α, and they are separated by closed taps, cf. Fig. 9.1. The total volume is $V = \sum_{\alpha=1}^{\nu} V_\alpha$. When the taps are opened the gases begin to mix, at constant p and T. Eventually we shall have a homogeneous mixture in the total volume. The partial pressures are p_α and we have $p = \sum_{\alpha=1}^{\nu} p_\alpha$. Since the gases are ideal, the total volume V is unchanged by mixing and so is the total internal energy U. This can easily be confirmed by use of the thermal and caloric equations of state. They read

before mixing: $p = N_\alpha \frac{1}{V_\alpha} kT$, $\quad U_\alpha = N_\alpha \left(z_\alpha k(T-T_R) + u_\alpha\left(T_R, p_R\right)\right)$

$$(16.1)$$

after mixing: $p_\alpha = N_\alpha \frac{1}{V} kT$, $\quad U = \sum_{\alpha=1}^{\nu} N_\alpha \left(z_\alpha k(T-T_R) + u_\alpha\left(T_R, p_R\right)\right).$

N_α is the number of particles of the gas α. z_α is a constant factor, equal to $\frac{3}{2}, \frac{5}{2}$, 3 respectively for one-, two-, or more-atomic gases. k is the Boltzmann constant and (T_R, p_R) are reference values of temperature and pressure, usually chosen as 298 K and 1 atm respectively.

The entropies of constituent α before and after mixing are

$$S_\alpha^B = N_\alpha \left(z_\alpha k \ln \frac{T}{T_R} + k \ln \frac{V_\alpha}{V_\alpha (T_R, p_R)} + s_\alpha (T_R, p_R) \right),$$

$$S_\alpha^A = N_\alpha \left(z_\alpha k \ln \frac{T}{T_R} + k \ln \frac{V}{V_\alpha (T_R, p_R)} + s_\alpha (T_R, p_R) \right).$$

(16.2)

Therefore the entropy of mixing $S_{\mathrm{Mix}} = \sum\limits_{\alpha=1}^{\nu} \left(S_\alpha^A - S_\alpha^B \right)$ is given by

$$S_{\mathrm{Mix}} = k \sum_{\alpha=1}^{\nu} N_\alpha \ln \frac{V}{V_\alpha} = k \sum_{\alpha=1}^{\nu} N_\alpha \ln \frac{N}{N_\alpha}.$$

(16.3)

Thus the entropy grows during mixing which is to be expected, since each constituent increases its volume.

By $(16.1)_1$ we have $N_\alpha = \frac{pV_\alpha}{kT}$ so that, given p, T and V_α, the particle number N_α is independent of the nature of the ideal gas and therefore S_{Mix} has the same value, irrespective of the gases which are brought to the mixing process. Now, suppose all volumes V_α contain the same gas; by $(16.3)_2$ that should not affect the entropy of mixing, although, of course, in that case the situations before and after opening the taps are identical, so that no mixing occurs. Surely this is a paradox, if there ever was one. It is called the *Gibbs paradox.*

To our knowledge the Gibbs paradox was never resolved, even though it has attracted much attention. Usually commentators – after explaining the paradox – veer round to statistical mechanics where there exists something which might be called a pseudo-Gibbs-paradox. This, however, is only superficially similar and, in fact, was quite satisfactorily explained after the advent of quantum mechanics. We proceed to discuss it.

16.2 The pseudo-Gibbs-paradox

We recall Sect. 3.6 with the reformulation of Boltzmann's entropy (3.20), namely

$$S = -k \int_V \int_{-\infty}^{\infty} \ln \frac{f}{b} f \, \mathrm{d}\mathbf{c} \mathrm{d}\mathbf{x}$$

(16.4)

into the equivalent forms (3.34), (3.35), which read with $b = eXY$

$$S = -k \sum_{\mathbf{xc}}^{PQ} \left(N_{\mathbf{xc}} \ln N_{\mathbf{xc}} - N_{\mathbf{xc}} \right) \quad \text{and} \quad S = k \ln \frac{1}{\prod\limits_{\mathbf{xc}}^{PQ} N_{\mathbf{xc}}!}.$$

(16.5)

We have explained – in Sect. 3.7 – that it is very tempting to smuggle a term $k \ln N!$ into $(16.5)_2$ so that the entropy may be written as

$$S = k \ln W \quad \text{with} \quad W = \frac{N!}{\prod\limits_{xc}^{PQ} N_{xc}!} . \tag{16.6}$$

The attraction of that expression for W is that it represents the number of possibilities to realize the distribution $\{N_{xc}\}$.

Recall that $(16.5)_2$ and $(16.6)_2$ are valid only, if N_{xc} are large numbers. We have argued before, however, cf. Sect.3.6, that these numbers may all be smaller than 1. Later in this chapter we shall see that this is indeed the case. But the pseudo-Gibbs-paradox belongs to the classical theory, where people *assumed* N_{xc} as big.

However, the arbitrary act of adding $k \ln N!$ to (16.5) costs dearly. The embellished entropy (16.6) is *not additive*. Let us consider:
We calculate W in (16.6) for a gas of N atoms in the volume V and in equilibrium so that the atoms are distributed homogeneously in space. In that case we write by use of the Stirling formula

$$S = k \left(N \ln N - N - \sum\limits_{xc}^{PQ} (N_{xc} \ln N_{xc} - N_{xc}) \right) .$$

P is the number of positions in V and Q is the number of velocities in $-\infty < c < \infty$. We assume that P is proportional to V and set $P = XV$. Homogeneity in space implies

$$N_{xc} = \frac{N_c}{P} \quad \forall \ \mathbf{x} \in V,$$

where N_c is the number of atoms with velocity $\mathbf{c}$. Thus we obtain

$$S = k \left(N \ln XV - \sum\limits_{c}^{Q} N_c \ln \frac{N_c}{N} \right) . \tag{16.7}$$

If we double the system, taking twice the number of atoms − and twice the number of atoms with each $\mathbf{c}$− in twice the volume, the entropy is not doubled according to (16.7). Therefore the entropy (16.6) is not additive; there is an additional term $k2N \ln 2$. We call this phenomenon the *pseudo-Gibbs-paradox*.
The pseudo-Gibbs-paradox is simply a punishment for the arbitrary addition of $k \ln N!$ to the true Boltzmann entropy (16.5). Indeed, when we apply the preceding chain of steps to (16.5) − instead of (16.6) − we obtain

$$S = k \left(N \ln X \frac{V}{N} + N - \sum\limits_{c}^{Q} N_c \ln \frac{N_c}{N} \right) . \tag{16.8}$$

This expression *is* additive; it doubles when N, N_c and V are doubled.
Therefore we may say that the pseudo-Gibbs-paradox is resolved. It does not represent a real problem but is rather a consequence of an overzealous extrapolation of Boltzmann's result (16.5). And the pseudo-paradox has nothing to do with the true paradox. Indeed, the true paradox about the entropy of mixing in (16.3) persists, irrespective of whether we calculate the entropy of mixing with entropies (16.7) or (16.8).

16.3 Boltzmann's entropy $(16.5)_1$ as $S = k \ln W$

Objective

After the entropy (16.6) has been discredited thoroughly by its non-additivity, it remains to be shown that the true entropy $(16.5)_1$ can be written as $S = k \ln W$ with W being the number of realizations of a distribution, albeit a different distribution from the one in (16.6). This problem was solved after the advent of quantum mechanics. We speak of *degenerate gases* when quantum effects are important, which happens for dense gases and gases at low temperatures.

Quantum corrections

We shall now proceed to show that Boltzmann's entropy $(16.5)_1$ *can indeed be written* as $k \ln W$, even without arbitrarily adding a term $k \ln N!$ to it. However it took quantum statistics to see that.

Among all the profound differences between classical mechanics and quantum mechanics there are only two which are relevant to the discussion of a monatomic gas:

- **There is no way to distinguish between identical particles**
  The classical idea is that we may mark the particles, e.g. "paint" them in different colors. But this is not only impractical but incompatible with quantum mechanics. In quantum mechanics a particle is characterized only by its mechanical state, that is to say be its position and momentum.
- **There are two types of particles, fermions and bosons**
  No two fermions may occupy the same state, but there is no such restriction on bosons; they may all pile up in one state.
  For a unified treatment of fermions and bosons we assume here that each state may be occupied by d particles. Of course $d = 1$ holds for fermions and $d = \infty$ for bosons and these are the only two cases that seem to occur in nature.

Number W of realizations and entropy of an element $d c d x$

As a result of these observations we must redefine the concepts of a realization of a distribution and of the distribution itself. In the classical theory a possible realization of a distribution $\{N_{\mathbf{x}\mathbf{c}}\}$ was that the atoms 1 through N_1 sat on place 1, the following N_2 atoms sat on place 2, etc. This involves identification of single atoms which is no longer permissible. Therefore we have to rethink the whole argument and to define a new distribution and new realizations.

We concentrate on the infinitesimal element $d c d x$ at $(\mathbf{x}, \mathbf{c})$ in $(\mathbf{x}, \mathbf{c})$-space, cf. Fig. 3.1, where we have

$$P_{d\mathbf{x}} Q_{d\mathbf{c}} = XY \, d c d x \quad - \text{ No. of points in } d c d x \tag{16.9}$$

$$N_{d\mathbf{x}d\mathbf{c}} = \overset{P_{d\mathbf{x}} Q_{d\mathbf{c}}}{\sum} N_{\mathbf{x}\mathbf{c}} = f(\mathbf{x}, \mathbf{c}) d c d x \quad - \text{ No. of atoms in } d c d x \tag{16.10}$$

The new distribution in the element dcdx is given by the set

$$\{p_l^{\mathbf{xc}}\} = \{p_0^{\mathbf{xc}}, p_1^{\mathbf{xc}}, ...p_d^{\mathbf{xc}}\} \tag{16.11}$$

which represents the number of points which are occupied by $0,1,...d$ atoms. Obviously the values $p_l^{\mathbf{xc}}$ must satisfy the constraints

$$\sum_{l=0}^{d} p_l^{\mathbf{xc}} = P_{\mathbf{dx}}Q_{\mathbf{dc}} \quad \text{and} \quad \sum_{l=0}^{d} l p_l^{\mathbf{xc}} = N_{\mathbf{dxdc}} . \tag{16.12}$$

A realization of this distribution is given by $\{N_{\mathbf{xc}}\}$, the number of atoms sitting on the points $(\mathbf{xc}) \in$ dcdx. [Ironically the new *realization* is the old *distribution*.]

By the rules of combinatorics the number of realizations for a distribution $\{p_l^{\mathbf{xc}}\}$ is equal to

$$W_{\mathbf{xc}} = \frac{(P_{\mathbf{dx}}Q_{\mathbf{dc}})!}{\displaystyle\prod_{l=0}^{d} p_l^{\mathbf{xc}}!}. \quad \text{Hence} \quad S_{\mathbf{xc}}\text{dcdx} = k \ln \frac{(P_{\mathbf{dx}}Q_{\mathbf{dc}})!}{\displaystyle\prod_{l=0}^{d} p_l^{\mathbf{xc}}!} \tag{16.13}$$

is the entropy of the atoms in the element dcdx and

$$S = k \ln \prod_{\mathbf{xc}} \frac{(P_{\mathbf{dx}}Q_{\mathbf{dc}})!}{\displaystyle\prod_{l=0}^{d} p_l^{\mathbf{xc}}!} \tag{16.14}$$

is the total entropy of the gas, where $\displaystyle\prod_{\mathbf{xc}}$ is the product over all elements dcdx.

Entropy in local equilibrium

It may be considered desirable to represent the entropy (16.13) of the element dcdx in terms of $N_{\mathbf{xc}}$, because that number is easier to interpret and more suggestive than $p_l^{\mathbf{xc}}$. Moreover $N_{\mathbf{xc}}$ occurs in the Boltzmann expression (16.5) which we wish to end up with as a special case of the present treatment.

For fermions there is no problem, because there is a one-to-one relation between $p_0^{\mathbf{xc}}$, $p_1^{\mathbf{xc}}$ — the only two types of points — and $P_{\mathbf{dx}}Q_{\mathbf{dc}}$ and $N_{\mathbf{dxdc}}$. Indeed from the constraints (16.12) we have

$$p_0^{\mathbf{xc}} = P_{\mathbf{dx}}Q_{\mathbf{dc}} - N_{\mathbf{dxdc}} \quad \text{and} \quad p_1^{\mathbf{xc}} = N_{\mathbf{dxdc}} . \tag{16.15}$$

For fermions the entropy (16.3) may therefore be written in the form

$$S_{\mathbf{xc}}^{F}\text{dxdc} = -k \left[N_{\mathbf{dxdc}} \ln \frac{N_{\mathbf{dxdc}}}{P_{\mathbf{dx}}Q_{\mathbf{dc}}} + P_{\mathbf{dx}}Q_{\mathbf{dc}} \left(1 - \frac{N_{\mathbf{dxdc}}}{P_{\mathbf{dx}}Q_{\mathbf{dc}}} \right) \ln \left(1 - \frac{N_{\mathbf{dxdc}}}{P_{\mathbf{dx}}Q_{\mathbf{dc}}} \right) \right] , \tag{16.16}$$

where the Stirling formula has been used for $P_{dx}Q_{dc}$ and N_{dxdc} — not for $N_{xc}(!)$.

For any value $d \geq 2$ such a representation is not possible, since there are more numbers p_l^{xc} to replace than we have constraints. However, in *local equilibrium* the numbers $p_l^{xc}(l = 0,1...d)$ assume values that maximize $S_{xc}dxdc$ under the constraints (16.12). A short calculation identifies those values as

$$p_n^{xc}|_{LE} = P_{dx}Q_{dc}\frac{e^{\lambda_{xc}n}}{\sum\limits_{l=0}^{d} e^{\lambda_{xc}n}}, \quad \text{with } \frac{N_{dxdc}}{P_{dx}Q_{dc}} = \frac{\partial}{\partial\lambda_{xc}}\left(\ln\sum\limits_{l=0}^{d} e^{\lambda_{xc}l}\right). \quad (16.17)$$

λ_{xc} is a Lagrange multiplier that takes care of the constraint $(16.12)_2$; it has to be determined from $(16.17)_2$. This, however, cannot be done analytically for an arbitrary d. Insertion of $p_n^{xc}|_{LE}$ into the entropy (16.13) gives

$$S_{xc}|_{LE} dcdx = -k\left[-N_{dxdc}\,\lambda_{xc} + P_{dx}Q_{dc}\ln\left(\sum\limits_{l=0}^{d} e^{\lambda_{xc}l}\right)\right]. \quad (16.18)$$

For bosons with $d = \infty$ the sum in (16.17) represents an infinite geometric series and λ_{xc} *can* be calculated in terms of N_{dxdc} and $P_{dx}Q_{dc}$. We obtain

$$\lambda_{xc} = \frac{\dfrac{N_{dxdc}}{P_{dx}Q_{dc}}}{1+\dfrac{N_{dxdc}}{P_{dx}Q_{dc}}}$$

and insertion into (16.18) gives for bosons in local equilibrium

$$S_{xc}|_{LE} dcdx = -k\left[N_{dxdc}\ln\frac{N_{dxdc}}{P_{dx}Q_{dc}} - P_{dx}Q_{dc}\left(1+\frac{N_{dxdc}}{P_{dx}Q_{dc}}\right)\ln\left(1+\frac{N_{dxdc}}{P_{dx}Q_{dc}}\right)\right]. \quad (16.19)$$

Comparison of the entropies (16.16) and (16.19) for fermions and bosons respectively reveals an interesting similarity. The two expressions differ only in several signs so that we may combine the equations in the form

$$S_{xc}dcdx =$$

$$-k\left[N_{dxdc}\ln\frac{N_{dxdc}}{P_{dx}Q_{dc}} \pm P_{dx}Q_{dc}\left(1\mp\frac{N_{dxdc}}{P_{dx}Q_{dc}}\right)\ln\left(1\mp\frac{N_{dxdc}}{P_{dx}Q_{dc}}\right)\right]\begin{array}{l}\text{fermions}\\\text{bosons}\end{array}. \quad (16.20)$$

As indicated, the upper sign refers to fermions, the lower one to bosons. The latter one is restricted to local equilibrium.

The classical limit

In the classical limit there must not be any difference between fermions and bosons. This means that the $\pm$-terms in (16.20) cannot affect the classical entropy. That is the case for

$$\frac{N_{\mathrm{dxdc}}}{P_{\mathrm{dx}}Q_{\mathrm{dc}}} \ll 1, \text{ i.e. very sparse occupation of the element dcdx.} \quad (16.21)$$

In that limit (16.20) degenerates to

$$S_{\mathbf{xc}}\mathrm{dcdx} = -k\,N_{\mathrm{dxdc}}\ln\frac{N_{\mathrm{dxdc}}}{P_{\mathrm{dx}}Q_{\mathrm{dc}}} - N_{\mathrm{dxdc}} \, . \quad (16.22)$$

We recall that $N_{\mathrm{dxdc}} = \overset{P_{\mathrm{dx}}Q_{\mathrm{dc}}}{\sum} N_{\mathbf{xc}}$ holds and that $N_{\mathbf{xc}} = \frac{N_{\mathrm{dxdc}}}{P_{\mathrm{dx}}Q_{\mathrm{dc}}}$ holds in local equilibrium. Therefore (16.22) may be written as

$$S_{\mathbf{xc}}\mathrm{dcdx} = -k\overset{P_{\mathrm{dx}}Q_{\mathrm{dc}}}{\sum}(N_{\mathbf{xc}}\ln N_{\mathbf{xc}} - N_{\mathbf{xc}}) \quad (16.23)$$

which is a sum over $P_{\mathrm{dx}}Q_{\mathrm{dc}}$ equal terms. The total entropy comes out from summing over all $\mathbf{x} \in V$ and all $\mathbf{c}$ in $-\infty < \mathbf{c} < \infty$:

$$S = -k\overset{PQ}{\sum}(N_{\mathbf{xc}}\ln N_{\mathbf{xc}} - N_{\mathbf{xc}}) \, . \quad (16.24)$$

This is Boltzmann's entropy $(16.5)_1$ or, equivalently, (16.4).

It is eminently reasonable that sparse occupation defines the classical limit. In fact, (16.21) means that most positions in the element dxdc are empty and some are occupied by one atom while very few, if any, have a higher occupancy. Under such circumstances it makes no difference whether the atoms are of Bose or Fermi type: there are so few of them that an attempt at double occupancy is a rare occasion indeed.

16.4 Conclusion

We have shown that the entropy $(16.5)_1$ of Boltzmann and the classical limit (16.24) of the entropy (16.14) are identical. This is important to know, because of the interpretation of entropy. Indeed, Boltzmann could not relate his entropy $(16.5)_1$ to an expression of the form $k\ln W$ — with W being the number of possibilities to realize a distribution — but the entropy (16.24) has been derived from just such a form, viz. (16.14). To be sure, Boltzmann's contemporaries could not know of any W other than $N!/\overset{PQ}{\underset{\mathbf{xc}}{\prod}} N_{\mathbf{xc}}$, whereas (16.24) is based on "quantum counting" and W comes from

$$W = \prod_{\mathbf{xc}} \frac{(P_{\mathrm{dx}}Q_{\mathrm{dc}})!}{\prod\limits_{l=0}^{d} p_l^{\mathbf{xc}}!}, \quad \text{cf. (16.14)},$$

where $\prod\limits_{\mathbf{xc}}$ is the product over all elements dcdx.

What the quantum argument has further told us is that we should never have converted S from $(16.5)_1$ into the factorial expression $(16.5)_2$. Indeed, the classical limit, where $(16.5)_1$ and (16.24) agree, holds only when $N_{\mathbf{xc}} = \frac{N_{\mathrm{dxdc}}}{P_{\mathrm{dx}}Q_{\mathrm{dc}}}$ is much smaller than 1. Therefore it is out of the question to apply the Stirling formula to it. Of course, classical physicists did not know that.

Finally, let us ask, whether we have proved that expressions like

$$S = k \ln W \quad \text{with} \quad W = \frac{N!}{\prod\limits_i N_i!} \tag{16.25}$$

are always invalid? The answer is: No! Such an expression for entropy is valid in those cases where W in (16.25) is the number of realizations of the distribution $\{N_i\}$ in a system of N particles. What we have proved in this section is only that W is not given by an expression of the type $(16.25)_2$ in a monatomic gas.

16.5 Equilibrium properties of a degenerate gas

By (16.9), (16.10) and (16.20) the entropy of a degenerate gas may be written in terms of the distribution function $f(\mathbf{x}, \mathbf{c}, t)$

$$S = -k \int \left[\ln \frac{f}{XY} \pm \frac{XY}{f} \left(1 \mp \frac{f}{XY} \right) \ln \left(1 \mp \frac{f}{XY} \right) \right] f \, \mathrm{dcdx}. \tag{16.26}$$

The equilibrium distribution f_E results from maximizing this expression under the constraints

$$N = \int\limits_V \int\limits_{-\infty}^{\infty} f \, \mathrm{dcdx} \quad \text{and} \quad U = \int\limits_V \int\limits_{-\infty}^{\infty} \frac{\mu}{2} c^2 f \, \mathrm{dcdx}. \tag{16.27}$$

A simple calculation provides

$$f_E = XY \frac{1}{e^{\alpha + \beta \frac{\mu}{2} c^2} \pm 1}, \quad \begin{matrix} \text{fermions} \\ \text{bosons} \end{matrix} \tag{16.28}$$

where α and β are Lagrange multipliers that take care of the constraints (16.27). These Lagrange multipliers must be determined as functions of U and N by insertion of f_E into the constraints

$$N = XY \int\limits_V \int\limits_{-\infty}^{\infty} \frac{1}{e^{\alpha + \beta \frac{\mu}{2} c^2} \pm 1} \, \mathrm{dcdx}, \quad U = XY \int\limits_V \int\limits_{-\infty}^{\infty} \frac{\frac{\mu}{2} c^2}{e^{\alpha + \beta \frac{\mu}{2} c^2} \pm 1} \, \mathrm{dcdx}. \tag{16.29}$$

and the equilibrium entropy results from insertion of (16.28) into (16.26)

$$S_E = k \left[\alpha N + \beta U \pm XY \int_{-\infty}^{\infty} \ln \left(1 \pm e^{-(\alpha + \beta \frac{\mu}{2} c^2)} \right) dc \right]$$

$$S_E = k \left[\alpha N + \frac{5}{3} \beta U \right] . \tag{16.30}$$

Obviously the Lagrange multipliers α, β cannot be calculated explicitly from (16.29) as functions of U and N. Therefore we identify α and β by comparison of $dS_E(U, V, N)$ with the Gibbs equation

$$dS_E = \frac{1}{T} dU + \frac{p}{T} dV - \frac{g}{T} d(\mu N), \tag{16.31}$$

where g is the specific free enthalpy, or chemical potential, cf. Sect. 7.6. Thus we obtain from (16.29) through (16.31)

$$\beta = \frac{1}{kT}, \quad p = \frac{2}{3} \frac{U}{V}, \quad \frac{k}{\mu} \alpha = -\frac{g}{T} . \tag{16.32}$$

We shall not discuss these formulae at any length nor do we pursue the interesting subject of degeneracy which is adequately treated in most books on statistical thermodynamic. The case offers explicit evidence for the third law and the capitulation of entropy.

The one aspect which we do mention is the classical limit and its interpretation in terms of the de Broglie wave character of the atoms. We have already seen in Sect. 16.3 that − and why − the ± alternative in (16.20), hence (16.29), becomes unimportant in a rarefied gas at high temperatures. This happens when α in (16.29) becomes much bigger than 1 and (16.29)$_1$ implies

$$e^{-\alpha} = \frac{N}{V} \frac{1}{XY \left(\frac{2\pi kT}{\mu} \right)^{3/2}} \underset{\text{by (14.1), (14.2)}}{=} \frac{N}{V} \left(\frac{h}{\sqrt{2\pi \mu kT}} \right)^3 \ll 1 . \tag{16.33}$$

The basis of the cubic expression in (16.33)$_2$ is the most probable de Broglie wave length of an atom in a gas with temperature T. We conclude that, for the classical limit to prevail, the number density must be so small that the de Broglie wave lengths of the atoms do not overlap. It is obvious from (16.33) that a high temperature − i.e. small de Broglie wave length − is conducive to the classical limit.

16.6 Transition probabilities of fermions and bosons

The equilibrium distributions (16.28) acquire a certain interpretability by the following argument which concerns the transition probabilities in a collision between atoms with velocities c and c^1 which, after the collision, have velocities c^l and $c^{1'}$. We assume that the transition probability is of the form[1]

$$P_{c^l c^{1'} \to cc^1} = AN_{xc} \, N_{xc^1} \, (1 \mp N_{xc^l}) \, (1 \mp N_{xc^{1'}}), \begin{matrix} \text{fermions} \\ \text{bosons} \end{matrix}$$

(16.34)

so that it depends not only on the occupation numbers of the elements of the colliding atoms before the collision, but also on those numbers after the collision. A is some factor of proportionality. Thus the transition of fermions is less probable if the target element is well-occupied, while the transition of bosons into such an element is more probable. For the reverse transition we assume an analogous expression for the transition probability, viz.

$$P_{cc^1 \to c^l c^{1'}} = AN_{xc^l} \, N_{xc^{1'}} \, (1 \mp N_{xc}) \, (1 \mp N_{xc^1}), \begin{matrix} \text{fermions} \\ \text{bosons} \end{matrix}$$

(16.35)

with the same factor of proportionality. In equilibrium the probabilities for the forward and backward transition should be equal so that

$$\frac{N_{xc}}{1 \mp N_{xc}} \quad \text{is a collisional invariant.}$$

(16.36)

Therefore the expression (16.36) must be a linear combination of the collisional invariants mass and energy of the atoms and we write

$$\frac{N_{xc}}{1 \mp N_{xc}} = \alpha + \beta \frac{\mu}{2} c^2 \quad \text{hence} \quad N_{xc} = \frac{1}{e^{\alpha + \beta \frac{\mu}{2} c^2} \pm 1}, \begin{matrix} \text{fermions} \\ \text{bosons} \end{matrix}.$$

(16.37)

By reference to (16.9), (16.10) this expression agrees with (16.28).

Originally the foregoing argument was used by Einstein who interpreted the Planck formula of radiation in terms of the stimulated emission of photons; photons are bosons in the state of complete degeneration, i.e. with $\alpha = 0$. Many decades later the idea led to the development of a maser − microwave amplification by stimulated emission of radiation − and then the laser.

---

[1] Recall that $N_{xc} = \frac{N_{dxdc}}{P_{dxQdc}}$ is the mean number of atoms on xc. It may be smaller than 1 and for fermions its largest value is equal to 1.

Thermodynamics of irreversible processes (TIP)

Non-equilibrium thermodynamics is an extensive and successful field which, however, is not the subject of this book. Here we merely review - in the briefest possible manner - the Eckart theory which provides us with an explicit expression for the dissipative entropy source in a viscous, heat-conducting fluid. We need this in Chap. 18 for an estimate of dissipation in the context of the entropy increase of radiation. And once we have that expression, we use it to disprove the "principle of minimum entropy production", a popular misconception in irreversible thermodynamics.

17.1 Viscous heat-conducting fluids

The objective of irreversible thermodynamics of viscous heat-conducting fluids is the determination of the 5 fields of

$$\text{mass density } \rho(\mathbf{x},t), \text{ velocity } v_i(\mathbf{x},t), \text{ temperature } T(\mathbf{x},t) \qquad (17.1)$$

in all points of the fluid and at all times.

For the purpose we need field equations, and these are based upon the equations of balance of mechanics and thermodynamics, viz. the conservation laws of mass and momentum and the equation of balance of internal energy, the so-called first law of thermodynamics

$$\dot{\rho} + \rho \frac{\partial v_j}{\partial x_j} = 0$$

$$\rho \dot{v}_i - \frac{\partial t_{ij}}{\partial x_j} = 0 \qquad (17.2)$$

$$\rho \dot{u} + \frac{\partial q_j}{\partial x_j} = t_{ij} \frac{\partial v_i}{\partial x_j} .$$

While these are 5 equations, they are not field equations for the fields (17.1). Indeed, T does not even appear in the equations (17.2), and instead they contain new quantities, viz. the

$$\text{(symmetric) stress tensor } t_{ij}$$
$$\text{heat flux } \qquad\qquad q_i \qquad\qquad\qquad (17.3)$$
$$\text{specific internal energy } \quad u \; .$$

In order to close the system we must find relations between t_{ij}, q_i, u and the fields (17.1), so-called constitutive relations, or equations of state.

In thermodynamics of irreversible processes, a theory universally known as TIP, such relations are derived in a heuristic manner from an entropy inequality that is based upon the Gibbs equation of equilibrium thermodynamics

$$\dot{s} = \frac{1}{T}\left(\dot{u} - \frac{p}{\rho^2}\dot{\rho}\right). \qquad (17.4)$$

s is the specific entropy. u and the pressure p are considered to be functions of ρ and T as prescribed by the caloric and thermal equations of state of equilibrium thermodynamics.[1] Elimination of $\dot{u}$ and $\dot{\rho}$ between (17.4) and $(17.2)_{1,3}$ and some rearrangement leads to the equation

$$\rho\dot{s} + \frac{\partial}{\partial x_i}\left(\frac{q_i}{T}\right) = q_i\frac{\partial\frac{1}{T}}{\partial x_i} + \frac{1}{T}t_{\langle ij\rangle}\frac{\partial v_{\langle i}}{\partial x_{j\rangle}} + \frac{1}{T}\left(\frac{1}{3}t_{ii} + p\right)\frac{\partial v_i}{\partial x_i} \qquad (17.5)$$

which may be interpreted as an equation of balance of entropy. That interpretation implies that

$$\varphi_i = \frac{q_i}{T} \qquad\qquad\qquad\qquad \text{is the entropy flux}^2,$$
$$(17.6)$$

$$\sigma_{mm} = -\frac{q_i}{T^2}\frac{\partial T}{\partial x_i} + \frac{t_{\langle ij\rangle}}{T}\frac{\partial v_{\langle i}}{\partial x_{j\rangle}} + \frac{1}{T}\left(\frac{1}{3}t_{ii} + p\right)\frac{\partial v_n}{\partial x_n} \quad \begin{matrix}\text{is the dissipative}\\\text{source density.}\end{matrix}$$

Angular brackets characterize symmetric traceless tensors, i.e. deviators. Inspection shows that the entropy source is a sum of products of

---

[1] This transfer of the equations of state from equilibrium thermodynamics to TIP is often called the "principle of local equilibrium". See also Sect. 7.3.

[2] Neither the assumption of local equilibrium nor the form $(17.6)_1$ of the entropy flux are fully confirmed by the kinetic theory of gases, cf. Sect. 3.5.

thermodynamic fluxes   and **thermodynamic forces**

heat flux q_i	temperature gradient $\dfrac{\partial T}{\partial x_i}$
stress deviator $t_{\langle ij \rangle}$	deviatoric velocity gradient $\dfrac{\partial v_{\langle i}}{\partial x_{j \rangle}}$ (17.7)
dynamic pressure $\pi = -\dfrac{1}{3}t_{ii} - p$	divergence of velocity $\dfrac{\partial v_n}{\partial x_n}$.

The dissipative entropy source must be non-negative. Assuming only linear relations between forces and fluxes TIP ensures the non-negative entropy source
by constitutive equations — phenomenological equations in the jargon of TIP
— of the type

$$q_i = -\kappa \frac{\partial T}{\partial x_i} \quad \kappa \geq 0$$

$$t_{\langle ij \rangle} = 2\mu \frac{\partial v_{\langle i}}{\partial x_{j \rangle}} \quad \mu \geq 0 \qquad (17.8)$$

$$\pi = -\lambda \frac{\partial v_n}{\partial x_n} \quad \lambda \geq 0 \,.$$

Along with the thermal and caloric equations of state $p = p(\rho, T)$, $u = u(\rho, T)$
the equations (17.8) are the constitutive equations of TIP. They are known as
the laws of Fourier and Navier-Stokes, with κ being the thermal conductivity
and μ and λ the shear and bulk viscosity respectively. All of these coefficients
may be functions of ρ and T. Of course, Fourier and Navier-Stokes did not
know the Gibbs equation nor did they need it. They proposed the equations
(17.8) on the basis of observation of the phenomena of heat conduction and
shear flow.

Inspection of $(17.6)_2$ shows that there are two "mechanisms" of entropy
source or dissipation, namely heat-conduction and viscous friction.

If the *transport coefficients* κ, μ and λ are known as functions of ρ and T,
the elimination of q_i and t_{ij} between (17.2) and (17.8) furnishes explicit dif-
ferential equations for the determination of the fields $\rho(\mathbf{x}, t)$, $v_i(\mathbf{x}, t)$, $T(\mathbf{x}, t)$.
Appropriate initial and boundary conditions must be chosen before a solution
can be attempted.

17.2 One-dimensional stationary heat conduction in a fluid at rest

In general the equations for ρ, v_i, T are strongly coupled non-linear partial differential equations. We simplify this system by assuming a stationary process

of heat conduction in the x-direction with the fluid at rest and with a homogeneous density. In that case the balance equations of mass and momentum are identically satisfied and the energy balance reduces to

$$\frac{\mathrm{d}q}{\mathrm{d}x} = 0 \quad \text{with} \quad q = -\kappa(T)\frac{\mathrm{d}T}{\mathrm{d}x}. \tag{17.9}$$

Hence follows an ordinary differential equation for $T(x)$, viz.

$$\frac{\mathrm{d}^2 T}{\mathrm{d}x^2} + \frac{\mathrm{d}\ln\kappa(T)}{\mathrm{d}T}\left(\frac{\mathrm{d}T}{\mathrm{d}x}\right)^2 = 0. \tag{17.10}$$

If $\kappa(T)$ were known, this equation could be solved and the solution could be fitted to the boundary values, e.g. the values of T on the two sides of the fluid.

17.3 Minimum dissipative entropy source

The above derivation of (17.10) could have been given by Fourier and every physicists since his time could have repeated it in half a minute. Nobody doubts that the result is correct as long as Fourier's law holds.

And yet, in the 1960's a principle of minimum dissipative entropy source could find adherents among physicists although it predicts a different result. That principle states that stationary fields — here the temperature field — are such that the entropy source is minimal. Let us investigate that proposition for the case considered in Sect. 17.2.

By (17.6)$_2$ and Fourier's law (17.9)$_2$ the entropy source in a fluid at rest between $x = 0$ and $x = L$ reads

$$\Sigma_{mm} = \int_0^L \kappa(T)\frac{1}{T^2}\left(\frac{\mathrm{d}T}{\mathrm{d}x}\right)\mathrm{d}x . \tag{17.11}$$

If the boundary temperatures are kept fixed, the Euler-Lagrange equation for a minimum of Σ_{mm} reads

$$\frac{\mathrm{d}^2 T}{\mathrm{d}x^2} + \frac{\mathrm{d}\ln(\frac{\sqrt{\kappa}}{T})}{\mathrm{d}T}\left(\frac{\mathrm{d}T}{\mathrm{d}x}\right)^2 = 0. \tag{17.12}$$

This equation is in contradiction with the energy balance (17.9) unless $\kappa \sim \frac{1}{T^2}$ holds, but no such case exists to our knowledge. Therefore we can safely discard the principle of minimum entropy source: It contradicts the *first law of thermodynamics*.

This does not mean that the principle of minimum entropy source has lost its adherents, nor have they concluded that the first law is wrong. The capacity for self-deception seems to be unlimited even among scientists, particularly if they have come in contact with thermodynamics.

Radiation thermodynamics

The thermodynamic densities and fluxes of a radiation field are best calculated by considering the field as a photon gas and using the methods of the kinetic theory of gases, cf. Chap. 3. Even without specifying the collision term — due to the interaction of matter and radiation — one may obtain important results about thermal and viscous dissipation in matter, if the radiative fluxes are known, and when stationary conditions prevail. This is so because in the stationary state the dissipative entropy source is balanced by the in- and outgoing fluxes of entropy.

This chapter prepares the reader for the subsequent one in which the dissipation of the earth's atmosphere is considered.

18.1 Photon transport equation and equations of transfer for radiation

Radiation consists of electromagnetic waves which in some ways — which physicists understand — behave like particles. Thus a wave of frequency ω and wave vector $\mathbf{k}$ corresponds to a photon of energy $\hbar\omega$ and momentum $\hbar k$.[1] The speed of the wave is c, the speed of light, so that the frequency ω and the wave number $k = |\mathbf{k}|$ are related by the dispersion relation $\omega = ck$. The unit vector $\mathbf{n}$ in the direction of propagation is given by $\mathbf{k}/k$.

A radiation field may be considered as a gas of photons. The number of photons in that gas with wave vectors between $\mathbf{k}$ and $\mathbf{k} + \mathbf{dk}$ is given by the phase density f. We have

$$\text{density of photons} = f(\mathbf{x}, \mathbf{k}, t)\mathbf{dk}. \tag{18.1}$$

We deal only with radiation in interaction with matter at rest.

In the space spanned by the coordinates of $\mathbf{x}$ and $\mathbf{k}$, the density f obeys a continuity equation with a production term, *the photon transport equation*

---

[1] $\hbar$ is the Planck constant divided by 2π.

$$\frac{\partial f}{\partial t} + cn_k \frac{\partial f}{\partial x_k} = \mathcal{S}(f); \tag{18.2}$$

$\mathcal{S}$ is the density of production of photons which is due to absorption, emission, and scattering. All three phenomena occur only by interaction of photons with matter, so that $\mathcal{S}$ is equal to zero when the space is free of matter.

The equation of transfer for radiation results from (18.2) by multiplication with a generic function $\psi(\mathbf{x}, t, \mathbf{k})$ and integration over $\mathbf{k}$ from $-\infty$ to $+\infty$. In this manner we obtain

$$\frac{\partial \int \psi f \mathrm{dk}}{\partial t} + \frac{\partial \int \psi c \frac{k_i}{k} f \mathrm{dk}}{\partial x_i} = \int \psi \mathcal{S} \mathrm{dk}. \tag{18.3}$$

With $\psi = 1$, $\hbar k_j$, $\hbar ck$, and $\psi = -k_B \left\{ \ln \frac{f}{y} - \left(1 + \frac{y}{f}\right) \ln \left(1 + \frac{f}{y}\right) \right\}$ we thus obtain equations of balance of the number of photons and of radiative momentum, energy and entropy. The appropriate densities, fluxes and sources can be read off from Table 18.1, where the Boltzmann constant is now denoted by k_B since the letter k is used for the wave number.

Table 18.1 Thermodynamic fields of radiation.

	density	flux	source density σ_{rm}
number	$n = \int f \mathrm{dk}$	$J_i = \int c\frac{k_i}{k} f \mathrm{dk}$	$\sigma_{rm}^n = \int \mathcal{S} \mathrm{dk}$
momentum	$p_j = \int \hbar k_j f \mathrm{dk}$	$P_{ij} = \int \hbar c \frac{k_i}{k} k_j f \mathrm{dk}$	$\sigma_{rm}^{p_j} = \int \hbar k_j \mathcal{S} \mathrm{dk}$
energy	$e = \int \hbar ck f \mathrm{dk}$	$Q_i = \int \hbar c^2 k_i f \mathrm{dk}$	$\sigma_{rm}^e = -r = \int \hbar ck \mathcal{S} \mathrm{dk}$
entropy	$h = -k_B \int [*] f \mathrm{dk}$	$\varphi_i = -k_B \int c\frac{k_i}{k}[*] f \mathrm{dk}$	$\sigma_{rm}^s = +k_B \int \ln \left(1 + \frac{y}{f}\right) \mathcal{S} \mathrm{dk}$

$$* = \ln \frac{f}{y} - \left(1 + \frac{y}{f}\right) \ln \left(1 + \frac{f}{y}\right)$$

Most of the entries in the table are self-explanatory. y is equal to $2/(2\pi)^3$ such that the smallest phase space element equals $\frac{1}{2}h^3$; the entropic terms are those appropriate for a Bose gas, cf. Sect. 16.5.[2] Inspection shows that the trace P_{ii} of the momentum flux equals the energy density. We may express this by saying that the radiation pressure $p = \frac{1}{3}P_{ii}$ equals 1/3 of the energy density

$$p = \frac{1}{3}e. \tag{18.4}$$

σ_{rm}^{pj} denotes the source of momentum of radiation due to matter and $\sigma_{rm}^{e} = -r$ denotes the source of energy of radiation due to matter. These notations have been chosen in accordance with Table A.1 so that the momentum and energy of matter and radiation together are conserved, if there is no gravitation. Recall also that the matter is considered at rest.

18.2 Planck distribution for equilibrium and equilibrium properties

For equilibrium we maximize the entropy under the constraint of fixed energy and obtain the Planck distribution

$$f_E(\mathbf{k}) = \frac{y}{e^{\beta \hbar c k} - 1}, \tag{18.5}$$

where β is the Lagrange multiplier that takes care of the constraint. Insertion into the densities of energy and entropy provides, cf. Table 18.1

$$e = \frac{1}{\pi^2}\frac{1}{\hbar^3 c^3}\frac{1}{\beta^4}\int_0^\infty \frac{x^3 \mathrm{d}x}{e^x - 1} \quad \text{and} \quad h = \frac{4}{3}k_B \frac{1}{\pi^2}\frac{1}{\hbar^3 c^3}\frac{1}{\beta^3}\int_0^\infty \frac{x^3 \mathrm{d}x}{e^x - 1} \tag{18.6}$$

Although we can solve $(18.6)_1$ so as to obtain β in terms of e, we should like to relate β to temperature. For that purpose we consider a volume V which contains the radiation energy eV and the radiation entropy hV. We may then use (18.6) to write

$$\mathrm{d}(hV) = k_B \beta(\mathrm{d}(eV) + p\mathrm{d}V). \tag{18.7}$$

Comparison with the Gibbs equation allows us to identify β

$$\beta = \frac{1}{k_B T} \tag{18.8}$$

and thus define the temperature of the photon gas. With (18.8) the Planck distribution (18.5) reads

$$f_E(\mathbf{k}) = \frac{y}{e^{\frac{\hbar c k}{k_B T}} - 1}. \tag{18.9}$$

---

[2] A photon is a Bose particle with two possible spins.

Table 18.2 gives equilibrium values for the radiative quantities in empty space. Most entries vanish, because of the isotropy of the equilibrium distribution and because there is no matter, so that $\mathcal{S}$ in (18.2) vanishes.

Table 18.2 Radiative fields in equilibrium with

$$a = \frac{1}{\pi^2}\frac{k_B^4}{\hbar^3 c^3}\int_0^{\infty}\frac{x^3\,\mathrm{d}x}{e^x - 1} = \frac{\pi^2}{15}\frac{k_B^4}{\hbar^3 c^3},$$

$\zeta(3) = 1.202$ is a value of the ζ-function.

	density	flux	source σ_{rm}
number	$\frac{15\zeta(3)}{\pi^4}\frac{a}{k_B}T^3$	0	0
momentum	0	$\frac{1}{3}aT^4\delta_{ij}$	0
energy	aT^4	0	0
entropy	$\frac{4}{3}aT^3$	0	0

18.3 Radiative fields in empty space in the neighbourhood of a spherical source

We refer to Fig. 18.1, which shows a spherical source S of radiation. Inside the source the radiation is supposed to be in equilibrium and the temperature is T_S. Therefore one concludes that in a point outside the source the distribution function is given by

$$f(\mathbf{x}, t, \mathbf{k}) = \begin{bmatrix} f_E(k; T_S) \text{ for } 0 \leq \varphi \leq 2\pi \;\; 0 \leq \beta \leq \beta_0 = \arcsin\frac{r}{R} \\ 0 \qquad\qquad \text{else} \end{bmatrix} \tag{18.10}$$

This distribution is strongly non-homogeneous and non-isotropic; it is therefore not an equilibrium distribution. We calculate the radiative fields of Table 18.1 and the results are summarized in Table 18.3.

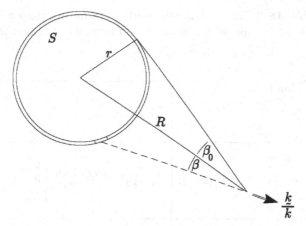

Fig. 18.1 On radiation at the distance R from
the center of sperical source of radius r.

The source densities σ_{rm} in Table 18.3 are all zero, because we are in empty space. To be sure the really interesting radiative phenomena are the interactions between radiation and matter which lead to sources σ_{rm} of all radiative quantities. In order to determine that production we need to know S, the phase density of photon production. We do not wish to go into that subject in detail. Rather we take a shortcut and calculate the sources σ_{rm} from our knowledge of the in- and outgoing fluxes in a stationary process. We shall do that in the next section.

The best known quantity in Table 18.3 is the energy flux Q_i. Its relation to the temperature T_S of the source is known as the *Stefan-Boltzmann law*, or sometimes the *fourth power law*. The law was found experimentally be Stefan, and Boltzmann put it into a thermodynamic setting, although not in the manner we have described. [After all Boltzmann did not know about photons and the Planck distribution, at least not when he worked on radiation.]

We may apply the law to the sun and the earth. The radiative energy flux reaching the sun from the earth can be measured to have the value $Q_\odot = 1341 \frac{W}{m^2}$. This value is known as the *solar constant*. With $R = 150 \cdot 10^9$m as the radius of the earth's orbit and $r_\odot = 0.7 \cdot 10^9$m as the radius of the sun we thus obtain the temperature of the sun as $T_\odot \approx 5740$ K.

Table 18.3 Thermodynamic fields of radiation from a
spherical source of radius r and temperature T_S.
$\sqrt{}$ stands for $\sqrt{1-\frac{r^2}{R^2}}$

	density	flux	source σ_{rm}
number	$n=\frac{1}{2}\frac{a}{k_B}\frac{15\zeta(3)}{\pi^4}T_S^3(2-\sqrt{})$	$J_i=\frac{c}{4}\frac{a}{k_B}\frac{15\zeta(3)}{\pi^4}T_S^3\frac{r^2}{R^2}\begin{bmatrix}0\\0\\1\end{bmatrix}$	0
momentum	$p_j=\frac{1}{4c}aT_S^4\frac{r^2}{R^2}\begin{bmatrix}0\\0\\1\end{bmatrix}$	$P_{ij}=\frac{1}{2}aT_S^4$ $\begin{bmatrix}\frac{1}{6}\sqrt{}^3-\frac{1}{2}\sqrt{}+\frac{1}{3} & 0 & 0\\0 & \frac{1}{6}\sqrt{}^3-\frac{1}{2}\sqrt{}+\frac{1}{3} & 0\\0 & 0 & \frac{1}{3}-\frac{1}{3}\sqrt{}^3\end{bmatrix}$	0
energy	$e=\frac{1}{2}aT_S^4\left(1-\sqrt{}\right)$	$Q_i=\frac{c}{4}aT_S^4\frac{r^2}{R^2}\begin{bmatrix}0\\0\\1\end{bmatrix}$	0
entropy	$h=\frac{1}{3}a\frac{4}{3}T_S^3\left(1-\sqrt{}\right)$	$\varphi_i=\frac{c}{4}a\frac{4}{3}T_S^3\frac{r^2}{R^2}\begin{bmatrix}0\\0\\1\end{bmatrix}$	0

Note that the energy and entropy fluxes in the immediate neighbourhood
of the source of radiation are given by

$$Q_i = \frac{c}{4}aT_S^4\begin{pmatrix}0\\0\\1\end{pmatrix} \quad \text{and} \quad \varphi_i = \frac{4}{3}\frac{c}{4}aT_S^3\begin{pmatrix}0\\0\\1\end{pmatrix} \tag{18.11}$$

respectively. These must therefore also be the values of the fluxes emanating
from a plane surface perpendicular to the direction $(0,0,1)$.

18.4 A comparison of the dissipative and the radiative entropy sources

We consider a block of concrete of thickness $L = 0.1$m and area A exposed to the solar radiation. The block absorbs the radiation in a thin surface layer. A part of the absorbed energy is emitted and the rest is transmitted through the block by heat flux. On the dark side − away from the sun − the block emits radiation according to its temperature and the fourth power law. The emitted radiation comes from a thin superficial layer on the dark side. Once the temperature has become a stationary field we denote the temperatures by T_1 on the sunny side and by T_2 on the dark side. There is no motion and the heat flux is governed by Fourier's law $q = -\kappa \frac{T_1 - T_2}{L}$, so that the temperature field inside the block is linear

$$T(x) = T_2 + \frac{T_1 - T_2}{L} x \qquad 0 \leq x \leq L . \qquad (18.12)$$

We balance the in- and outgoing energies in the whole block and in the surface layer on the dark side. Thus we obtain respectively

$$Q_\odot - \frac{c}{4}aT_1^4 = \frac{c}{4}aT_2^4 \quad \text{and} \quad \frac{c}{4}aT_2^4 = \kappa \frac{T_1 - T_2}{L}. \qquad (18.13)$$

These are two equations for T_1 and T_2. With $\kappa = 0.74 \frac{\text{W}}{\text{mK}}$, appropriate for concrete and $Q_\odot = 1341 \frac{\text{W}}{\text{m}^2}$ we obtain as the only relevant solution

$$T_1 = 355\text{K} \quad \text{and} \quad T_2 = 296\text{K}. \qquad (18.14)$$

Thus we have calculated the temperature field in the block.

Next we calculate the entropy source Σ^s occuring in the block. In principle there are four additive contributions, viz.

- Σ^s_{rr} − entropy source of radiation due to the interaction of photons among themselves. Such an interaction does not occur under the conditions considered here so that Σ^s_{rr} is always 0 for us.
- Σ^s_{rm} − entropy source of radiation due to the interaction of radiation with matter. By Table 18.1 we have

$$\Sigma^s_{rm} = k_B \int_V \int \ln\left(1 + \frac{y}{f}\right) S \mathrm{d}k \mathrm{d}x. \qquad (18.15)$$

This source occurs in the superficial layers, where the radiation is absorbed and emitted and its calculation requires detailed knowledge of the production rate S. In the absence of that knowledge Σ^s_{rm} may be calculated − at least under stationary conditions − as the difference of the out- and in-going entropy fluxes. According to Table 18.3 we thus have

$$\Sigma^s_{rm} = \left[\frac{c}{4}a\frac{4}{3}T_1^3 - \frac{c}{4}a\frac{4}{3}T_\odot^3\frac{r_\odot^2}{R^2} + \frac{c}{4}a\frac{4}{3}T_2^3\right]A = 5.032\ A\frac{\text{W}}{\text{m}^2\text{K}}. \qquad (18.16)$$

- Σ^s_{mr} – entropy source of matter due to the interaction of matter with radiation. According to Table A.1 this contribution is given by

$$\Sigma^s_{mr} = \int \frac{r}{T} dV, \qquad (18.17)$$

where r is the source density of radiative energy. Since in our case r occurs only in the thin superficial layers of the block at temperatures T_1 and T_2 we may calculate Σ^s_{mr} from the in- and outgoing radiative energy fluxes as

$$\Sigma^s_{mr} = \left[\frac{1}{T_1} \left(Q_\odot - \frac{c}{4} a T_1^4 \right) - \frac{1}{T_2} \frac{c}{4} a T_2^4 \right] A = -0.243 \, A \frac{W}{m^2 K}. \qquad (18.18)$$

- Σ^s_{mm} – dissipative entropy source of matter due to heat conduction. The dissipative entropy source density σ^s_{mm} is given by (17.6) which in the present case of no motion reduces to $\frac{\kappa}{T^2} \left(\frac{dT}{dx} \right)^2$. By integration over x we obtain

$$\Sigma^s_{mm} = \frac{\kappa}{L} \frac{(T_1 - T_2)^2}{T_1 T_2} A = 0.243 A \frac{W}{m^2 K}. \qquad (18.19)$$

We note that the entropy source $\Sigma^s_{mr} + \Sigma^s_{mm}$ of matter is zero, because whatever entropy is produced by heat conduction is balanced by a decrease of entropy of matter due to absorption and emission of radiation. Therefore we may calculate the entropy source due to heat conduction – more generally: due to heat conduction and viscous friction – from the absorbed and emitted energy fluxes of radiation. This is useful to know when the dissipation due to conduction and friction cannot be calculated explicitly because the temperature field and the flow field are too complicated. Thus for the earth's atmosphere such a calculation would seem to be impossible, while the radiative energy fluxes are known and so are the temperatures where absorption and emission occurs, cf. Chap. 19. Therefore in such cases the dissipation may still be determined provided stationary conditions prevail.

Because of the foregoing observation the total entropy source Σ^s – the sum of all three terms Σ^s_{rm}, Σ^s_{mr}, Σ^s_{mm} – is entirely a source of entropy of radiation. And that source is positive, because more entropy of radiation is produced by emission than is lost by absorption.

Finally we remark that the entropy source Σ^s_{rm} of radiation is more than twenty times bigger than the dissipative entropy source Σ^s_{mm} of heat conduction, cf. (18.16) and (18.19).

18.5 A radiation temperature

Since r and $\frac{r}{T}$ are the radiative source densities of internal energy and material entropy respectively, and since $-r$ is the source denstiy of radiative energy, it is tempting to define a radiative temperature Ξ by setting the source density of radiative entropy equal to $\sigma_{rm}^s = -\frac{r}{\Xi}$. In the case of two parallel plates of stationary temperatures T_1 and T_2, where the interaction of matter and radiation occurs only in thin surface layers we obtain by use of (18.11)

$$\Xi = \frac{3}{4}\frac{T_1^4 - T_2^4}{T_1^3 - T_2^3} \cdot \tag{18.20}$$

That "radiative temperature" depends on the temperatures of *both* plates and *it has the same value in both plates*. When the two plates have nearly the same temperature T, the radiative temperature is equal to T because, according to (18.20), that is the limiting value of Ξ for $T_1 \to T_2$.

If matter and radiation interact at some depth, rather than only near the surface, we may still define the radiative temperature field, but things are more complicated in that case and we have to know the phase density S of production of photons in (18.2). The special case of absorption and emission only – without scattering – has been dealt with by Müller.

19

Dissipative entropy source of the earth

A bare planet or satellite, like mercury or the moon, have a fairly trivial balance of incoming and outgoing radiation. That balance determines the surface temperature and the entropy sources. It is always true − as it was for the concrete block of Chap. 18 − that the entropy source of radiation due to matter is between one and two orders of magnitude bigger than the dissipative entropy source due to heat conduction and viscous friction.

The earth's climate is determined by the incoming solar radiation and by the efflux of radiation emanating from the clouds and from the earth's surface. Meteorologists have determined, how these fluxes are partitioned and what the temperatures of the atmospheric layers are. We use these data to determine the dissipative entropy source in the atmosphere and compare it to the dissipation of entropy by mankind. The latter is numerically negligible.

19.1 Absorption coefficient, albedo, and emission coefficient

In Chap. 18 we have tacitly assumed that all the radiation that falls on a body is absorbed. This is often not the case, because usually not all frequencies *can* be "handled" by the atoms of a body and the radiation of those frequencies is reflected. Thus the earth reflects 28% of the incoming solar radiation, preferably the short wave lengths, which is why our planet appears blue if viewed from space. The fraction of radiation absorbed by the earth is $a_E = 0.72$ and that number is called the *absorption coefficient*; $1 - \alpha_E = 0.28$ is called the *albedo* of the earth.

Also in Chap. 18 we have assumed that a body of temperature T emits the fluxes $\frac{c}{4}aT^4$ of energy and $\frac{c}{4}a\frac{4}{3}T^3$ of entropy. This again is not generally so, because some bodies can only emit certain frequencies or radiation in a certain range − or in certain ranges − of frequency. This is the case, for instance, for the ozone layer in the stratosphere, which emits only radiation at or near the wave length $9.5 \cdot 10^{-6}$m and the radiative flux is only 7.6% of what the stratosphere might emit if it were capable of emitting all frequencies. Therefore the fluxes of energy and entropy emanating radially from the stratosphere are

$$\varepsilon_s \frac{c}{4} a T_s^4 \quad \text{and} \quad \varepsilon_s \frac{c}{4} a \frac{4}{3} T_s^3 \tag{19.1}$$

respectively, where $\varepsilon_s = 0.076$ is the *emission coefficient* of the stratosphere.

These two concessions to the real world — the albedo of the earth and the emission coefficient of the stratosphere — are imperative, if we wish that our treatment of the earth's atmosphere be at least roughly quantitative. Other radiative emissions, like the one from the surfaces of the clouds or from the litho-/hydrosphere, are taken to be ideal in the sense that the emission coefficient equals 1. A correct treatment with measured radiation spectra is quite difficult and is far beyond the scope of this book.

19.2 Bare planets and the bare earth

Bare planets with homogeneous temperature

The simplest model for estimating planetary temperatures ignores the atmospheres and the temperature difference between day and night. What is not ignored, however, is the albedo, the fraction of reflected radiation which, according to Sect. 19.1 has the value 0.28 for the earth.[1] By Table 18.3 the incoming radiative fluxes of energy and entropy are therefore given by

$$\alpha \frac{c}{4} a T_\odot^4 \frac{r_\odot^2}{R^2} \pi \rho^2 \quad \text{and} \quad \alpha \frac{c}{4} a \frac{4}{3} T_\odot^3 \frac{r_\odot^2}{R^2} \pi \rho^2 \tag{19.2}$$

respectively, where R is the distance of the planet from the sun and ρ is the radius of the planet. α is the absorption coefficient. For earth it has the value 0.72. The corresponding outgoing fluxes are, according to (18.11)

$$\frac{c}{4} a T_P^4 4 \pi \rho^2 \quad \text{and} \quad \frac{4}{3} \frac{c}{4} a T_P^3 4 \pi \rho^2 \tag{19.3}$$

respectively, where T_P represents the temperature of the planet.

We assume stationary conditions so that the energy fluxes $(19.2)_1$ and $(19.3)_1$ must be balanced. Thus we obtain an equation for T_P given the albedo and the radius R of the planetary orbit. For the earth with $\alpha_E = 0.72$ and $R_E = 150 \cdot 10^9$m we obtain

$$T_E = 255 \text{ K.} \tag{19.4}$$

---

[1] It is true that the lack of atmosphere changes the albedo, but we ignore that aspect for the present calculation.

Mars and Jupiter are considerably colder because of their greater distance from the sun. With $\alpha_M = 0.85$ and $\alpha_J = 0.48$ and the appropriate values of R we obtain

$$T_M = 215 \text{ K} \qquad\qquad T_J = 101 \text{ K} . \qquad\qquad (19.5)$$

Planetary temperatures so calculated are sometimes called astronomical temperatures. They are too low for planets with an atmosphere, because of the green-house-effect, cf. later sections of this chapter.

As in Chap. 18 for the model case of the concrete block we decompose the entropy source of the planet into three parts, namely Σ^s_{rm}, Σ^s_{mr}, Σ^s_{mm}. And we calculate those parts just as we did for the block: By balancing fluxes.

Thus for the bare earth we calculate Σ^s_{rm}, the entropy source of radiation by interaction with matter, from the difference of the out- and in-going entropy fluxes $(19.3)_2$ and $(19.2)_2$. For $T = T_E = 255$ K we obtain[2]

$$\frac{\Sigma^s_{rm}}{4\pi\rho^2} = 1.1975 \frac{\text{W}}{\text{m}^2\text{K}} . \qquad\qquad (19.6)$$

Once again we conclude — just like for the concrete block of Chap. 18 — that the entropy source Σ^s_{rm} is positive since more entropy of radiation is produced by emission than is lost by absorption.

The material entropy source $\Sigma^s_{mr} + \Sigma^s_{mm}$ is still zero, of course, since there is no material entropy flux in or out of the planet. But Σ^s_{mr} is also zero, since the absorption and emission of radiation both occur at $T_E = 255$ K in a thin layer at the earth's surface. Therefore we have

$$\Sigma^s_{mr} = \int_V \frac{r}{T} dV = -\frac{1}{T_E}\left(-\alpha\frac{c}{4}aT_\odot^4\left(\frac{r_\odot}{R}\right)^2\pi\rho^2 + \frac{c}{4}aT_E^4 4\pi\rho^2\right) = 0, \quad (19.7)$$

and it follows that Σ^s_{mm} also vanishes. The reason is that there is no heat conduction, since the temperature field was assumed homogeneous.

Bare earth with day/night temperature differences

These results change when we allow for a difference of temperature of the two half spheres of the earth during day and night. We take the temperatures to be $\Delta T = 5$ K above and below T_E and calculate T_E from the equation

$$-\alpha_E\frac{c}{4}aT_\odot^4\left(\frac{r_\odot}{R}\right)^2\pi\rho^2 + \frac{c}{4}a\left(T_E + \Delta T\right)^4 2\pi\rho^2 + \frac{c}{4}a\left(T_E - \Delta T\right)^4 2\pi\rho^2 = 0.$$

Obviously the influence of ΔT on T_E is of second order in $\frac{\Delta T}{T_E}$ and therefore the effect on T_E is small: We obtain

$$T_E = 254.8 \text{ K}$$

---

[2] We refer all Σ's in this chapter to a square meter of the earth's surface, since that is a convenient measure.

which is indeed only marginally different from (19.4). For the same reason we may take Σ_{rm}^s to have essentially the same value as before, viz. (19.6).

However, the day and night difference in temperature does affect the dissipative entropy source Σ_{mm}^s significantly, because it is no longer equal to zero now. To be sure $\Sigma_{mr}^s + \Sigma_{mm}^s$ is still zero but $\Sigma_{mr}^s = -\Sigma_{mm}^s$ is now given by

$$\Sigma_{mr}^s = \frac{\alpha_E \frac{c}{4} a T_\odot^4 \left(\frac{r_\odot}{R}\right)^2 \pi \rho^2}{T_E + \Delta T} - \frac{c}{4} a \left(T_E + \Delta T\right)^3 2\pi\rho^2 - \frac{c}{4} a \left(T_E - \Delta T\right)^3 2\pi\rho^2.$$

Hence follows with $\rho = 6.3 \cdot 10^6$ m for the radius of the earth

$$\frac{\Sigma_{mm}^s}{4\pi\rho^2} = 0.0152 \frac{\text{W}}{\text{m}^2\text{K}} . \tag{19.8}$$

We may consider this dissipative entropy source as due to heat conduction between the hot and cold part of the earth. In the concrete block, cf. Chap. 18, we were able to calculate such a term directly but for the earth this is difficult, if not impossible, because the temperature field is too complicated. Therefore we chose to calculate Σ_{mm}^s as $-\Sigma_{mr}^s$. In the sequel we shall always follow that path. The method is efficient as long as stationarity prevails and as long as we know the fluxes of energy and the temperatures where energy is absorbed and emitted.

We observe again — as in Chap. 18 — that the dissipative entropy source Σ_{mm}^s is only a small fraction of the radiative one, viz. Σ_{rm}^s. In the present case the ratio is 1.3% as a comparison between (19.6) and (19.8) shows.

19.3 The earth with an atmosphere. Energy fluxes and temperatures

We recall the solar constant $Q_\odot = 1341\frac{\text{W}}{\text{m}^2}$ and conclude that the solar radiative power reaching the earth is $P_\odot = Q_\odot \pi \rho^2$, where ρ is the earth's radius as before. We also recall the albedo $1 - \alpha_E = 0.28$ of the earth which represents the fraction of $P_\odot$ which is reflected. The reflection is largely due to the atmosphere and we may therefore say that the solar power penetrating the atmosphere equals $0.72P_\odot$. In this section and the next all partitions of power and all contributions to heating and cooling — whether radiative are otherwise — are measured in terms of $P_\odot$. We ignore day/night differences for simplicity.

Meteorologists have observed, measured, or reasonably assumed that the radiative power $0.72P_\odot$ penetrating the atmosphere may be split into three parts, cf. Fig. 19.1.

$$\begin{aligned} P_\odot^{LH} &= 0.47\, P_\odot - \text{absorbed by the litho-hydrosphere} \cdot \\ P_\odot^T &= 0.22\, P_\odot - \text{absorbed by the troposphere} \\ P_\odot^S &= 0.03\, P_\odot - \text{absorbed by the stratosphere.} \end{aligned} \tag{19.9}$$

On the other hand, the outgoing components of power are estimated to be

$$P^{LH} = 0.05 \ P_\odot \ - \text{emitted into space by the litho-hydrosphere}$$

$$P^T \ = 0.64 \ P_\odot \ - \text{emitted into space by the troposphere} \tag{19.10}$$

$$P^S \ = 0.03 \ P_\odot \ - \text{emitted into space by the stratosphere.}$$

The sums of the incoming and outgoing powers are equal of course under stationary conditions. Both are equal to $0.72 P_\odot$.

These contributions are schematically represented by arrows in Fig. 19.1. The decomposition of $P_\odot$ illustrated in the figure is taken from the book by Kleemann & Meliß cited in the figure caption. We have supplemented their representation with temperatures and heights. The two tropospheric heights, which are exhibited in the figure, we consider as the heights of the upper and lower surfaces of the cloud cover. We proceed to explain.

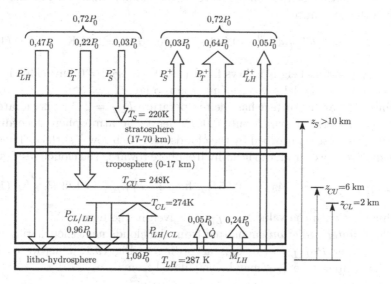

Fig. 19.1 Energy fluxes in the atmosphere.
According to Kleeman, M., Meliß, M. (1988),
Regenerative Energiequellen, Springer Heidelberg.

In order to fix the ideas we assume that the cloud cover of the earth is homogeneous. It lets 47% of the solar radiative power $P_\odot$ pass through, cf. (19.9). But only 5% of $P_\odot$ passes the cloud cover in the reverse direction emanating from the earth's surface. The reason is that the incoming radiation — emanating from the sun at 5740 K — has a different frequency spectrum than the outgoing radiation, cf. the Planck distribution (18.9). The sun's infrared radiation penetrates clouds more efficiently than the earth's surface radiation.

And yet, according to Fig. 19.1, there is a considerable absorption of solar radiation in the troposphere, where the clouds are, viz. 22%. And most of the emission of earth radiation — namely 0.64 $P_\odot$ emanates from the clouds or — more properly — from the upper layer of the clouds.

The stratosphere, on the other hand, reemits what it absorbs in the ozone layer.

Meteorologists have invented a *standard atmosphere* which is characterized by the ground temperature $T_{LH} = 287$ K and a temperature gradient $\gamma = \frac{0.65 \text{ K}}{100 \text{ m}}$ up to a height of 10 km. Above that height the temperature is constant. Thus the temperature field of the standard atmosphere is given by

$$T(z) = T_{LH} - \gamma z \qquad 0 < z < 10 \text{ km} \qquad (19.11)$$

That observation along with the information contained in Fig. 19.1 permits the determination of the temperature of the clouds' surfaces. For the upper cloud surface we have

$$\frac{c}{4} a T_{CU}^4 4\pi\rho^2 = 0.64 P_\odot, \text{ hence } T_{CU} = 248 \text{ K}. \qquad (19.12)$$

The corresponding height follows from (19.11) as $z_{CU} = 6$ km. [This is what our atmospheric model says; z_{CU} may be too large.]

Since the earth's surface has the temperature $T_{LH} = 287$ K, it radiates the power $\frac{c}{4} a T_{LH}^4 4\pi\rho^2$ of which only $0.05 P_\odot$ leave the atmosphere according to Fig. 19.1; the rest is absorbed by the clouds near their lower surface. Therefore the radiative power from the litho-hydrosphere towards the clouds results from

$$P_{LH/CL} = \frac{c}{4} a T_{LH}^4 4\pi\rho^2 - 0.05 P_\odot, \text{ hence } P_{LH/CL} = 1.09 P_\odot. \qquad (19.13)$$

which agrees with the value for $P_{LH/CL}$ given in the figure.

The *counter radiation* emanating from the clouds near their lower surface has the power $P_{CL/LH} = 0.96 P_\odot$ according to Fig. 19.1. Thus we conclude that the temperature T_{CL} may be calculated from

$$\frac{c}{4} a T_{CL}^4 4\pi\rho^2 = 0.96 P_\odot, \text{ hence } T_{CL} = 274 \text{ K}. \qquad (19.14)$$

By (19.11) this corresponds to the height $z_{CL} = 2$ km.

The temperature T_s of the stratosphere cannot be calculated so easily, because the emission coefficient $\varepsilon_s = 0.076$ must be taken into account, cf. Sect. 19.1. Therefore we have from the data of Fig. 19.1

$$\varepsilon_s \frac{c}{4} a T_S^4 4\pi\rho^2 = 0.03 P_\odot, \text{ hence } T_S = 220 \text{ K}. \qquad (19.15)$$

This is the value of T_s listed in the figure. According to the temperature profile of the standard atmosphere T_s is first reached at the height $z = 10$ km.

So much for the incoming and outgoing radiative power. There is also heating $\dot{Q}$ of the troposphere by the lithosphere which amounts to $\dot{Q} = 0.05P_\odot$ according to Fig. 19.1. $\dot{Q}$ is due to the thermal contact of the air with the ground.

Finally there is energy lost from the litho-hydrosphere due to the evaporation of water. In Fig. 19.1 that contribution is denoted by M_{LH} and it amounts to $0.24P_\odot$. It is informative, perhaps, to relate M_{LH} to the rate $\dot{m}$ of evaporating mass. That mass rises by thermal convection and condenses — and freezes — at the proper height to form the clouds. Eventually it falls down as water or snow. On the rise this motion leaves the earth's surface with the energy rate $\dot{m}h''(T_{LH})$ and in falling it returns an energy rate equal to $\dot{m}h'(T_{LH})$ to the surface, where h'' and h' are the specific enthalpies of water vapour and liquid water respectively. We assume that both flows have the temperature $T_{LH} = 287\mathrm{K}$ on the surface. Therefore we may set

$$M_{LH} = \dot{m}\left(h''(T_{LH}) - h'(T_{LH})\right) = \dot{m}r(T_{LH}) = 0.24P_\odot, \qquad (19.16)$$

where $r(T_{LH}) = 2.47 \cdot 10^3 \frac{\mathrm{kJ}}{\mathrm{kg}}$ is the specific heat of evaporation. From $(19.16)_3$ we obtain for the mass rate of evaporation

$$\dot{m} = 16.25 \cdot 10^6 \frac{t}{s}. \qquad (19.17)$$

Of particular interest for the climate is the so-called green-house effect. In a greenhouse the incoming solar radiation is essentially unaffected by the glass roof and is absorbed by the ground. The radiation from the ground upwards is, however, stopped by the glass and comes back from there as "counter radiation" appropriate to the temperature of the roof. In the case of the atmosphere the glass roof is represented by the cloud layer, of course, and the counter radiation of power $0.96P_\odot$ keeps our ground temperature at the comfortable value of 287K.

19.4 Dissipative entropy source of the troposphere

The atmosphere is strongly dissipative because of heat conduction and viscous heating which are the two "mechanisms" of dissipative entropy source according to Chap. 17.

The direct calculation of

$$\Sigma_{mm}^s = \int\limits_V \left(-q_i \frac{\partial \frac{1}{T}}{\partial x_i} + \frac{t_{ij}}{T} \frac{\partial v_i}{\partial x_j} \right) dV \qquad (19.18)$$

by integration over the volume of the atmosphere is, however, quite impossible because we do not know the details of the fields of temperature and velocity in the atmosphere. Therefore we proceed indirectly as follows.

We use the entropy balance of matter and the energy balance of radiation, viz. according to Table A.1 and Table 18.1

$$\int_V \frac{\partial \rho s}{\partial t} dV + \int_{\partial V} \left(\rho s v_i + \frac{q_i}{T} \right) n_i dA = \int_V \frac{r}{T} dV + \Sigma_{mm}^s$$

$$\int_V \frac{\partial e}{\partial t} dV + \int_{\partial V} Q_i n_i dA = -\int_V r dV \quad . \tag{19.19}$$

We assume stationarity and apply these equations to the atmospheric model described in Sec. 19.3. The model is characterized by four thin layers of volumes V_α and surfaces ∂V_α– with $\alpha = 1$ through 4 – having temperatures T_α and radiative energy sources $r_\alpha V_\alpha$. The layers form the parts of the lithohydrosphere, of the clouds, and of the stratosphere where radiation is absorbed and emitted.

First of all we apply the entropy balance $(19.19)_1$ to a sphere containing the whole earth and its atmosphere. Thus we obtain, since on that sphere there is no material entropy flux $\rho s v_i + q_i/T$,

$$0 = \sum_{\alpha=1}^{4} \frac{r_\alpha V_\alpha}{T_\alpha} + \Sigma_{mm}^s \quad , \tag{19.20}$$

where Σ_{mm}^s is the total dissipative entropy source of the earth's atmosphere. Also we apply $(19.19)_2$ to the layers α individually thus obtaining

$$\int_{\partial V_\alpha} Q_i n_i dA = -r_\alpha V_\alpha \qquad \alpha = 1, 2, 3, 4. \tag{19.21}$$

Elimination of $r_\alpha V_\alpha$ between these equations provides Σ_{mm}^s in terms of the radiative energy fluxes, viz.

$$\Sigma_{mm}^s = \sum_{\alpha=1}^{4} \frac{1}{T_\alpha} \int_{\partial V_\alpha} Q_i n_i dA \quad . \tag{19.22}$$

From Fig. 19.1 we read off the fluxes and obtain

$$\Sigma_{mm}^s = \frac{P_\odot}{T_{LH}} (-0.47 - 0.96 + 1.09 + 0.05) +$$

$$+ \frac{P_\odot}{T_{CL}} (0.96 - 1.09) + \frac{P_\odot}{T_{CU}} (-0.22 + 0.64) + \tag{19.23}$$

$$+ \frac{P_\odot}{T_s} (-0.03 + 0.03)$$

$$\frac{\Sigma_{mm}^s}{4\pi\rho^2} = 0.0699 \frac{W}{m^2 K} .$$

The last equation provides the entropy production per m^2 of the earth's surface, because that is a convenient measure.

In this model the dissipative entropy source is entirely due to the tropo-sphere, since the stratosphere has no entropy source; it merely absorbs and emits the same energy at the same temperature. If we allowed day/night tem-perature differences there would be a small contribution due to the strato-sphere. Also, of course, the contribution of the troposphere would change slightly. Weiss has calculated the effect of the day/night differences.

Nor does this model allow for a contribution of the litho-hydrosphere to Σ_{mm}^s, because it does not specify conduction or convection in the earth or in the oceans.

Even so, it is interesting to compare the dissipative entropy source (19.23) of the troposphere to the dissipative entropy source due to mankind. We burn coal, oil, natural gas and nuclear fuel and use hydroelectric energy at a rate of $1.270 \cdot 10^{13}$W, and we produce energy from biomass, geothermal sources, wind and sun at a rate of $0.0084 \cdot 10^{13}$ W. It is reasonable to assume that this power is eventually dissipated at the temperature of our environment, which is 287 K. Therefore the dissipated entropy production due to mankind equals $\Sigma_{mm}^{s(human)} = 4.4 \cdot 10^{10} \frac{W}{K}$ or, if we refer this to the surface of the earth

$$\frac{\Sigma_{mm}^{s(human)}}{4\pi\rho^2} = 8.8 \cdot 10^{-5} \frac{W}{m^2 K} . \tag{19.24}$$

Comparison with (19.23) shows that this is a little more than $1^o/oo$ of the dissipative entropy source of the earth.

So, our efforts are like nothing compared to hurricanes, or thunderstorms, or the weather as such.

However, environmentalists need not worry that we are undermining their prediction of doom and human self destruction. After all, what is $1^o/oo$? As interest rate for our money in the bank $1^o/oo$ is very little, but it is a fair amount, if it represents the alcohol in our blood.

19.5 Entropy source Σ_{rm}^s of radiation

We use Table 18.1 to write the entropy balance of radiation in the form

$$\int_V \frac{\partial h}{\partial t} dV + \int_{\partial V} \varphi_i n_i dA = \Sigma_{rm}^s . \tag{19.25}$$

We apply this balance to a concentric sphere surrounding the earth and atmo-sphere and, as always, we assume stationary conditions. The influx of entropy into that sphere is given by $(19.2)_2$ with $\alpha = 0.72$ and the outfluxes are given by $(18.11)_2$ or, in the case of the stratosphere by $(19.1)_2$. By Fig. 19.1 there are three outfluxes: one each for the temperatures T_S and T_{CU} and T_{LH}. The latter contributes only the fraction $\frac{0.05}{1.14}$ to the outflux of entropy, since the fraction $\frac{1.09}{1.14}$ is stopped by the clouds according to Fig. 19.1.

Therefore we have

$$\frac{\Sigma^s_{rm}}{4\pi\rho^2} = -0.72\frac{4}{3}\frac{c}{4}T_\odot^3\frac{r_\odot^3}{R^2}\frac{1}{4}+$$

$$+0.076\frac{4}{3}\frac{c}{4}aT_S^3 + \frac{4}{3}\frac{c}{4}aT_{CU}^3 + \frac{0.05}{1.14}\frac{4}{3}\frac{c}{4}aT_{LH}^3 \qquad (19.26)$$

$$= 1.2366\frac{W}{m^2K}.$$

Once again we conclude — as we did for the concrete block in Chap. 18 — that the entropy source of radiation is an order of magnitude bigger than the dissipative entropy source (19.23).

20

Socio-thermodynamics — integration and segregation in a population

A population of metaphorical hawks and doves, competing for the same resource is a well-known model of game theory characterized by a certain contest strategy. This model is modified here by making rewards and penalties dependent on the prevailing price level. Also a second strategy is invented which the hawks and doves may choose or not depending on which strategy provides a higher gain.

It turns out that at intermediate and high price levels the population will not achieve maximum gain, if hawks and doves remain homogeneously mixed. Rather they will partly or even fully segregate to maximize their gain.

The ideas presented here are extrapolations of thermodynamics of binary mixtures whose components may mix at high temperature while at low temperature they exhibit miscibility gaps.

20.1 Maximization of gain

Strategies

We consider a mixed population of N birds — zN hawks and $(1-z)N$ doves — who feed on the same resource. The resource may either be collected or bought. Its monetary value is 5τ. τ itself is the price of some standard good and it may vary in time. We shall refer to τ as the *price*. The price is out of control of the population; it may change, so that there are periods of high price and periods of low price.

Upon collecting the resource the birds compete and they may choose between two strategies:

Strategy A

If two hawks meet over the resource, they fight until one is injured. The winner gains the value 5τ while the loser, being injured, needs time for healing his wounds. Let that time be such that the hawk must buy 2 resources, worth 10τ, to feed himself during the convalescence. Two doves do not fight. They merely engage in a symbolic conflict, posturing and threatening, but not actually fighting. One of them will eventually win the resource – always with the value 5τ– but they have both lost time such that after every dove-dove encounter they need to satisfy their need by buying part of a resource, worth 2τ. If a hawk meets a dove, the dove walks away while the hawk wins the resource; there is no injury nor is any time lost.

Assuming winning and losing the fights or the posturing game equally probable we conclude that the expectation values for the gain per encounter are given by the arithmetic mean value of the gains in winning and loosing, i.e.

$$e_A^{HH} = \frac{1}{2}\left(5\tau - 10\tau\right), \quad e_A^{HD} = 5\tau, \quad e_A^{DH} = 0, \quad e_A^{DD} = \frac{1}{2}5\tau - 2\tau \quad (20.1)$$

for the four possible encounters HH, HD, DH, DD.

Note that by (20.1) both the fighting of the hawks and the posturing of the doves are irrational compulsive acts, or luxuries. Both species would do better, if they cut down on these activities or if they abandoned them altogether. That observation has let to the formulation of strategy B which, however, is subtly different from strategy A in some other features as well.

Strategy B

The hawks adjust the severity of the fighting – and thus the gravity of the injury – to the prevailing price. If the price τ of the standard good is higher than 1, they fight less, so that the time of convalescence is shorter and the value of resources to be bought during convalescence reduces from 10τ to $10\tau(1-0.3(\tau-1))$. Likewise the doves adjust the duration of the posturing so that the payment for lost time reduces from 2τ to $2\tau(1-0.3\,(\tau-1))$. But that is not all: In strategy B doves will still not fight when they find themselves competing with hawks, but they will try to grab the resource and run. Let them be successful 4 out of 10 times. However, successful or not, they risk injury from the enraged hawk and hence need a period of convalescence at a cost of $10\tau(1+0.8(\tau-1))$. Note that the severity of the injury and hence the expected cost of convalescence rises when the price grows beyond 1.

Thus the expected gains for the encounters e_B^{HH}, e_B^{HD}, e_B^{DH}, e_B^{DD} read

$$e_B^{HH} = \tfrac{1}{2}\left(5\tau - 10\tau\left(1 - 0.3\left(\tau - 1\right)\right)\right)$$

$$e_B^{HD} = \tfrac{6}{10}5\tau$$

$$e_B^{DH} = \tfrac{4}{10}5\tau - 10\tau\left(1 + 0.8\left(\tau - 1\right)\right)$$

(20.2)

$$e_B^{DD} = \tfrac{1}{2}5\tau - 2\tau\left(1 - 0.3\left(\tau - 1\right)\right).$$

The specific numbers in (20.1) and (20.2) have been chosen so as to fit our conception of likely behaviour of the species. $\tau = 1$ is a reference price for which both strategies agree — except for the "grab and run" feature. The numbers represent the rules of the game which the birds play. It is clear that fighting and posturing should be punished, not rewarded. Therefore we must limit τ to values smaller than 4.33.[1]

Expectation values

The expectation values for the gains of hawks and doves per encounter under the two strategies $i = (A, B)$ read

$$e_i^H = z\, e_i^{HH} + (1 - z)\, e_i^{HD} \quad \text{and} \quad e_i^D = z\, e_i^{DH} + (1 - z)\, e_i^{DD}, \qquad (20.3)$$

since the probabilities for meeting a hawk or a dove are equal to z and $(1 - z)$ respectively.

Finally the expected gain per encounter per bird reads

$$e_i = z\, e_i^H + (1 - z)\, e_i^D \qquad (i = A, B) . \qquad (20.4)$$

Obviously e_i for $i = A, B$ are both quadratic functions of z whose graphs are represented by parabolae. Figure 20.1 shows a tableau of five such pairs of parabolae which differ by the price.

---

[1] A condition like that could be avoided, if we allowed the penalties to depend on τ non-linearly. We do not do that for simplicity.

Maximum gain

In this section we take it for granted that the birds, given the choice of strategy, will choose *the* strategy that offers the higher gain. Inspection of Fig. 20.1 shows that for the low price $\tau = 0.5$ the high-gain-strategy is strategy A irrespective of the hawk fraction z, because the A-curve lies entirely above the B-curve. We expect the population to be homogeneously mixed for all z and employing strategy A. This is still the case for $\tau = 1$.

More interesting is the case $\tau \approx 2$, because the graph $max\ [e_A(z), e_B(z)]$ is non-concave. Therefore the concavification of the graphs – represented by the straight lines, tangent to $e_A(z)$, in Fig. 20.1c,d – provides a higher gain for the population.

We consider the situation shown in Fig. 20.1c: A dove-rich population with the hawk fraction $z = Z$ can increase its gain by decomposing into a pure-dove colony with $z = 0$ employing strategy B and a mixed colony with $z = z_L$ employing strategy A.

The actual gain is then given by

$$e(Z) = e_B(0) + \frac{e_A(z_L) - e_B(0)}{z_L} Z, \tag{20.5}$$

which is marked by a fat point in Fig. 20.1c. The gain $e(Z)$ is bigger than both $e_A(Z)$ and $e_B(Z)$ and that is why the population falls apart into colonies. Similar arguments hold for the same price $\tau = 2$ in a hawk-rich population under the right tangent which touches the curve $e_A(z)$ at z_R.

For the higher price $\tau = 2.8$ the two separate tangents of the curve $e_A(z)$ coalesce; the point of contact occurs at $z = z_E$, cf. Fig. 20.1d. This means that for all $z < z_E$ the population decomposes into pure-dove colonies $z = 0$ with strategy B and mixed colonies with $z = z_E$ with strategy A. For $z > z_E$ the pure colonies are pure-hawk $z = 1$.

A new situation occurs for $\tau > 2.8$, e.g. $\tau = 3$, cf. Fig. 20.1e. In that case the concavification of $max\ [e_A(z), e_B(z)]$ connects the end points of $e_B(z)$ by a straight line. Accordingly we expect a decomposition of the population into colonies of pure doves and pure hawks, both employing strategy B.

Strategy diagram

It is convenient to summarize the foregoing discussion in a (τ, z)-strategy diagram. In order to construct that diagram we project the straight parts of the concave envelopes for price τ onto the horizontal line $\tau = const.$ and connect their end points, cf. Fig. 20.1c and f. In this manner we obtain two triangular segments marked II and III which contain the (τ, z)-pairs for which pure-species colonies with strategy B coexist with mixed populations employing strategy A.

We identify the status of the populations characterized by the pairs (τ, z) inside the areas I through IV in Fig. 20.1f as follows.

I.   Fully integrated population of hawks and doves employing strategy A.
II. Coexisting colonies of pure doves with strategy B and of a hawk-dove mixture with strategy A.
III. Coexisting colonies of pure hawks with strategy B and of a hawk-dove mixture with strategy A.
IV. Coexisting colonies of pure doves and pure hawks both employing strategy B. Total segregation.

We conclude that, if a ruler in the kingdom of metaphorical birds wishes to maintain — for reasons of his own — an integrated population, he best keeps the price of resources low. Or if he cannot prevent the price from rising, he best has a population with the phase fraction z_E. However, there is nothing he can do, if he lets the price rise beyond $\tau = \tau_E = 2.8$. In that case segregation is inevitable.

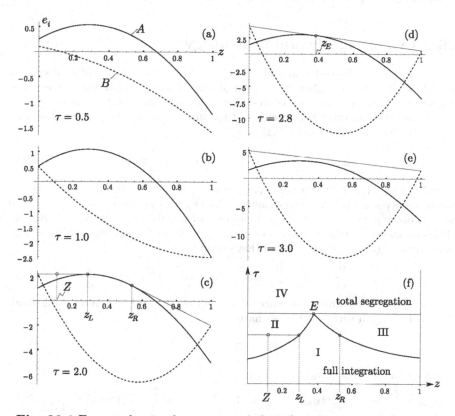

Fig. 20.1 Expected gains for strategy A (solid) and strategy B (dashed) for five different prices τ, as functions of the hawk fraction z, cf. (a) through (e). Also: strategy diagram in the (τ, z)-plane.

Analogy with thermodynamics

Chemists, physicists and engineers will recognize the construction and interpretation of the strategy diagram from metallurgy of alloys or physico-chemistry of solutions where phase diagrams are constructed in this manner, cf. Chap. 11. Total segregation of the constituents occurs in those fields as the phenomenon of unmixing, like when fat globules float on a watery soup.

Such analogies emphasize the point of view that physical or sociological systems of many individual elements have common properties, whether the elements are atoms and molecules or birds and – maybe – men.

And yet, the foregoing is not socio-*thermodynamics*. While it seems plausible that the birds strive for maximum gain, a principle like that is "begging the question". What we are missing sofar are the analogues in sociobiology to the first and second laws of thermodynamics. In order to obtain such a structure we have to endow the population of hawks and doves with more complexity. We proceed to do that in Sects. 20.3 ff.

20.2 Evolutionarily stable strategy

The hawk and dove population and strategy A with $\tau = 1$ are an often discussed model of game theory applied to the field of sociobiology. Sociobiologists have conceived the idea that a species has an evolutionary advantage if its expected gain is bigger than that of the other species so that the fraction of the privileged species grows. We have by (20.1) and (20.3)

$$e_A^H(z) = -\frac{15}{2}z + 5 \qquad e_A^D(z) = -\frac{1}{2}z + \frac{1}{2} \,. \qquad (20.6)$$

Therefore $e_A^H(z \gtrsim 0) = 5$ and $e_A^D(z \gtrsim 0) = \frac{1}{2}$ so that few hawks in a predominantly dove population have an advantage. Also $e_A^H(z \lesssim 1) = -\frac{5}{2}$ and $e_A^D(z \lesssim 1) = 0$ so that doves are better off in a hawk-dominated population. Both populations are therefore *evolutionarily unstable*. The strategy is an evolutionarily stable strategy only when $e_A^H(z) = e_A^D(z)$ holds and that occurs at

$$z_{ESS} = \frac{9}{14} \,, \text{ hence } e_A^H(z_{ESS}) = \frac{5}{28} \text{ and } e_A^D(z_{ESS}) = \frac{5}{28} \,. \qquad (20.7)$$

We may also ask where the expected gain e_A per bird per encounter is maximal. By (20.4) – again with $\tau = 1$ – we have

$$e_A(z) = -7\,z^2 + 4z + \frac{1}{2} \qquad (20.8)$$

which has a maximum for

$$z_{\text{Max}} = \frac{2}{7}, \text{ hence } e_A(z_{\text{Max}}) = \frac{15}{14} \text{ and } e_A^H(z_{\text{Max}}) = \frac{20}{7}, \ e_A^D(z_{\text{Max}}) = \frac{5}{14}. \qquad (20.9)$$

Both the hawks and the doves – and the "average bird" – are thus better off for $z = z_{\text{Max}}$ than for $z = z_{ESS}$ but, of course, the hawks are forty times better off and the doves are only twice as well off. The hawks will therefore undermine the population; they will grow at the expense of the doves and thereby hurt their own gain and that of the doves.

Our modification of the game, described in Sect. 20.1 does not allow the hawk fraction to change. The evolutionary "mechanism" is switched off and the richness of the model in our case results from the possibility of the birds to choose between strategies.

20.3 The first law of socio-thermodynamics

Motivation

Interesting as the foregoing considerations of game theory may be, they do not constitute socio-thermodynamics. The description of a thermodynamic binary mixture is not complete unless we allow for non-homogeneous fields of temperature, number density and concentration and unless we specify heating and working on the surface. Socio-thermodynamics – if it deserves that name – must provide equivalent or analogous notions and concepts. We proceed to suggest these in an intuitive and heuristic manner.

Temperature and price

Prices have a tendency to become equal across borders and boundaries, if trade is admitted. This is analogous to temperature which is continuous across a heat-conducting wall according to the zeroth law of thermodynamics. Therefore we identify price with temperature. In general the price τ is a non-homogeneous time-dependent field inside the volume V where the population lives, i. e. τ depends on the position $x \in V$ and on time t.

Number density and hawk concentration

Likewise the number density of birds is generally a non-homogeneous time-dependent field   denoted by $n(x, t)$, and so is the hawk fraction $z(x, t)$. The constant total number N of birds, and the constant total number N^H of hawks result from integration over V

$$N = \int_V n(\mathbf{x}, t)\mathrm{d}\mathbf{x}, \quad N^H = \int_V n(\mathbf{x}, t)z(\mathbf{x}, t)\mathrm{d}\mathbf{x} . \tag{20.10}$$

Energy and gain; shortfall

Energy "wants" to *decrease* as we have often argued before, but our methapho-rical birds want to *increase* the gain e. As an analogy to energy we therefore suggest to take the *shortfall* defined as

$$u = 5\tau - e \tag{20.11}$$

which decreases when e increases.

e is the expected gain per bird and encounter, as before, and 5τ is the value of the maximal possible gain; a bird would achieve that gain, if it succeeded in each encounter. Therefore u represents the shortfall of gain from the maximum.

Since by (20.1) through (20.4) e depends on τ and z – both functions of $x \in V$ and t in general – $u(x,t)$ is also a field and the total shortfall U result from integration

$$U = \int_V n(\mathbf{x}, t) u(\mathbf{x}, t) d\mathbf{x} \ . \tag{20.12}$$

At any point $\mathbf{x}$, or inside the volume element $d\mathbf{x}$ at $\mathbf{x}$, we assume that one strategy prevails at time t, either A or B. But at different points different strategies may be employed, or also at one point for different times.

Caloric equations of state

Insertion of e_i from (20.1) through (20.4) into (20.11) provides the "caloric equation of state", viz. the dependence of the shortfalls $u_i (i = A, B)$ on τ and z. We obtain

$$u_A = \{ \underbrace{\left[\tfrac{15}{2}\tau\right]}_{u_A^H(\tau)} z + \underbrace{\left[\tfrac{9}{2}\tau\right]}_{u_A^D(\tau)} (1 - z) \ \} - 7z\tau(1 - z) \tag{20.13}$$

$$u_B = \{ \underbrace{\left[9\tau - \tfrac{3}{2}\tau^2\right]}_{u_B^H(\tau)} z + \underbrace{\left[\tfrac{51}{10}\tau - \tfrac{3}{5}\tau^2\right]}_{u_B^D(\tau)} (1 - z) \ \} - \left(\tfrac{71}{10}\tau - \tfrac{101}{10}\tau^2\right) z(1 - z).$$

Both are increasing functions of τ in the admissible range $0 < \tau \leq 4.33$. [Thus the "specific heats" $\frac{\partial u_i}{\partial T}$ are positive.] The last terms in (20.13) – the ones with $z(1 - z)$ may be considered as shortfalls of mixing in the sense in which quantities of mixtures are related to those of pure constituents, cf. Sect. 9.1.

Heating and value transfer

The total shortfall U in V may change when the price changes due to trade across the surface, or also just the possibility of trade. We write

$$\left.\frac{dU}{dt}\right|_{\text{value transfer}} = \dot{Q} \tag{20.14}$$

and call $\dot{Q}$ the value transfer. It is akin to the heating of a thermodynamic system on the boundary which leads to a temperature change.

Working

Let the boundary ∂V or part of the boundary be movable under an external pressure. A volume decrease $\frac{dV}{dt} < 0$ is assumed to increase the shortfall $U -$ hence the price – inside V; after all, part of the habitat is lost and so are the resources that might be collected there. We set

$$\left.\frac{dU}{dt}\right|_{\text{working}} = -p_0 \frac{dV}{dt}, \tag{20.15}$$

where $p_0 > 0$ is called the external pressure. It is taken to be homogeneous on the movable parts of ∂V. The right hand side of (20.15) is called the working of the external pressure p_0.

First law

We combine (20.14) and (20.15) and arrive at the first law of socio-thermodynamics

$$\frac{dU}{dt} = \dot{Q} - p_0 \frac{dV}{dt} . \tag{20.16}$$

Population pressure

The external pressure p_0, which occurs in the 1st law, is not the only pressure occuring in socio-thermodynamics. Indeed, the birds, walking randomly in V, fill that volume whatever its size. Therefore they exert a *population pressure* p on the boundary ∂V, since that boundary prevents the expansion of the population into a bigger volume.

Thermal equation of state

For a single species and single-strategy-population it seems reasonable to assume that the population pressure p at some point near ∂V is proportional to the population density n of birds and to the price τ. Indeed, if n grows, we expect the push against the wall to grow and, if τ grows, we imagine the random walk of the birds to become more frantic so that the push against the wall becomes bigger. Thus we let the population pressure be given by

$$p = \alpha \, n \, \tau, \tag{20.17}$$

where $\alpha > 0$ is a factor of proportionality typical for a population.

The factor α may depend on the species and – for each species – on the strategy. Therefore we are looking at four possible values for α, viz. α_A^H, α_B^H, α_A^D and α_B^D. In subsequent calculations we choose the α's as shown in Table 20.1.

Table 20.1   Pressure coefficients α

	A	B
D	0.5	3.5
H	0.5	3.5

If hawks and doves are both present at $\mathbf{x}$ we consider the population pressure to be the sum

$$p_i = p_i^H + p_i^D = [\alpha_i^H z + \alpha_i^D (1 - z)] n\tau \qquad (i = A, B) . \qquad (20.18)$$

Recall that one strategy – either A or B – prevails locally.

The form (20.17) of the pressure has overtones of the ideal gas relation. There is an important difference, however, since α is not a universal constant in the present case.

First law for a quasistatic process

If value transfer and the working of the external pressure p_0 – both connected with the boundary ∂V – are applied rapidly, we expect the fields τ and n and z inside V to be strongly non-homogeneous and time-dependent. However, when value transfer and working stop, we assume that those fields will relax to homogeneity and stationarity. In that case we expect that the population pressure p and the external pressure p_0 are equal. We extrapolate this expectation by assuming that in a slow application of value transfer and working the fields will always be homogeneous and $p = p_0$ will prevail during the whole process. In that case (20.16) reduces to its quasistatic form

$$\frac{dU}{dt} = \dot{Q} - p\frac{dV}{dt} . \qquad (20.19)$$

Carnot process

Under quasistatic conditions – i.e. homogeneous fields $n(x)$, $z(x)$, $\tau(x)$ – U and p in (20.19) depend on τ and V. We may then define quasistatic processes $\tau(t)$, $V(t)$ and calculate the value transfer $\dot{Q}$ from (20.19). Or we may consider an "adiabatic" process with $\dot{Q} = 0$ and calculate $\tau(t)$ in terms of $V(t)$. In particular we may consider a Carnot cycle consisting of four branches, two adiabatic ones and two with constant prices. We do that for a single-strategy A population with the equations of state, cf. (20.18), (20.13)

$$p = [\alpha_A^H z + \alpha_A^H (1 - z)] \frac{N}{V}\tau \quad \text{and} \quad U = N \left(\frac{9}{2}\tau - 4\tau z + 7\tau z^2 \right) .$$

Thus we may calculate the work $A = -\int p\,dV$ and transfers $= \int \dot{Q}\,dt$ for all branches and then define an efficiency as the ratio of the total work $A_{\circlearrowleft}$ and the positive transfers. After a little calculation we obtain

$$e = 1 - \frac{\tau_L}{\tau_H} \tag{20.20}$$

in obvious analogy to the Carnot efficiency of an ideal gas. τ_L and τ_H are the low and high prices respectively.

It may seem facetious to talk about a Carnot process in a population, but to the thermodynamicist this is entirely plausible. In Chap. 1 we have seen the important role played by Carnot processes in the Clausius derivation of entropy. Here we shall duplicate that derivation and therefore we shall need to know (20.20).

Dimensionless variables

All quantities introduced here — the price, the pressure, the volume, the gain, the shortfall, etc — are dimensionless. This is a convenient convention. Everything else, like measuring the shortfall in a particular currency, or the population pressure in bar, would be inappropriate.

20.4 The second law of socio-thermodynamics

Formulation

We recall the second law in the Clausius formulation, cf. Chap.1

Heat cannot pass by itself from a cold to a hot body.

In analogy we formulate a second law of socio-thermodynamics

Value cannot pass by itself from a cheap to a dear population.

This is an eminently plausible extrapolation: Think of a weekend market with two stalls selling the same good at different prices. The cheap stall will sell quickly so that he comes to realize that he may raise the price and still sell all his goods. The dear stall sells little and he will therefore lower the price so as to sell more. Thus there is a transfer of value between the stalls from dear to cheap — *without any transfer of goods*. The opposite does not happen ever; or so our second law maintains.

If this law is accepted along with the first laws (20.16), (20.19), we find ourselves in the same position as Clausius, cf. Chap. 1. And, if we proceed in the manner pursued by Clausius we arrive at his result about entropy, viz.

$$\frac{dS}{dt} \geq \frac{\dot{Q}}{\tau_0} \tag{20.21}$$

τ_0 is the price at the boundary of the population where the value transfer $\dot{Q}$ into the population occurs. The equality holds for a quasistatic process. In such a process τ_0 equals τ, the homogeneous price. S is called the entropy of the population.

20.5 Gibbs equation and equations of state for the entropies s_A and s_B

In a quasistatic process, where the equality holds and the boundary temperature τ_0 equals the homogeneous temperature τ inside the population, we may eliminate $\dot{Q}$ between (20.19) and (20.21) and obtain the Gibbs equation

$$dS = \frac{1}{\tau}(dU + pdV) . \tag{20.22}$$

This equation permits the calculation of the entropies of the single-species-populations by insertion of U from (20.13) and p from (20.17) and integration. With $s = \frac{S}{N}$ we obtain

$$s_A^H(\tau,p) = s_A^H(\tau_R,p_R) + \left(\frac{15}{2} + \alpha_A^H\right) \ln\frac{\tau}{\tau_R} - \alpha_A^H \ln\frac{p}{p_R}$$

$$s_A^D(\tau,p) = s_A^D(\tau_R,p_R) + \left(\frac{9}{2} + \alpha_A^D\right) \ln\frac{\tau}{\tau_R} - \alpha_A^D \ln\frac{p}{p_R}$$

$$\tag{20.23}$$

$$s_B^H(\tau,p) = s_B^H(\tau_R,p_R) + \left(9 + \alpha_B^H\right) \ln\frac{\tau}{\tau_R} - 3\left(\tau - \tau_R\right) - \alpha_B^H \ln\frac{p}{p_R}$$

$$s_B^D(\tau,p) = s_B^D(\tau_R,p_R) + \left(\frac{51}{10} + \alpha_B^D\right) \ln\frac{\tau}{\tau_R} - \frac{6}{5}\left(\tau - \tau_R\right) - \alpha_B^D \ln\frac{p}{p_R}$$

τ_R and p_R are arbitrary reference values of price and population pressure, which we shall both choose to be equal to 1 in the sequel.

The integrability conditions implied by the Gibbs equation are trivially saisfied, since u_i ($i = A, B$) depend on τ only and since p is linear in τ.

In a homogeneous hawk-dove mixture under the pressure p the partial pressures are

$$p_i^H = \frac{\alpha_i^H z}{\alpha_i^H z + \alpha_i^D(1-z)}p, \quad \text{and} \quad p_i^D = \frac{\alpha_i^D(1-z)}{\alpha_i^H z + \alpha_i^D(1-z)}p .$$

Therefore the specific entropies of the hawk-dove mixtures at τ and p read for $i = A, B$

$$s_i(\tau, p, z) = \left\{ s_i^H(\tau, p) z + s_i^D(\tau, p)(1 - z) \right\} -$$

$$-\alpha_i^H z \ln \frac{\alpha_i^H z}{\alpha_i^H z + \alpha_i^D(1 - z)} -$$

$$-\alpha_i^D(1 - z) \ln \frac{\alpha_i^D(1 - z)}{\alpha_i^H z + \alpha_i^D(1 - z)}, \tag{20.24}$$

where the last two terms represent the entropies of mixing of the population.

20.6 Availability

Elimination of $\dot{Q}$ between the first and second laws for a rapid — i.e. non-quasistatic — process provides the inequality

$$\frac{\mathrm{d}(U - \tau_0 S + p_0 V)}{\mathrm{d}t} \leq -S \frac{\mathrm{d}\tau_0}{\mathrm{d}t} + V \frac{\mathrm{d}p_0}{\mathrm{d}t} . \tag{20.25}$$

It follows that — for fixed τ_0 and p_0 on the boundary — the availability

$$\mathfrak{A} = U - \tau_0 S + p_0 V \tag{20.26}$$

decreases. Thus when the stationary state — i.e. equilibrium — is reached, $\mathfrak{A}$ is minimal.

We write $\mathfrak{A}$ as an integral over the specific values u, s, and $v = \frac{1}{n}$ and the integration is over the total number of birds

$$\mathfrak{A} = \int_N (u(\mathbf{x}) - \tau_0\, s(\mathbf{x}) + p_0\, v(\mathbf{x}))\, \mathrm{d}N \longrightarrow \text{Minimum.} \tag{20.27}$$

We ask for the fields $\tau(\mathbf{x}), p(\mathbf{x}), z(\mathbf{x})$ which make this expression minimal under the constraint

$$N^H = \int_N z(\mathbf{x}) \mathrm{d}N . \tag{20.28}$$

From here on we follow exactly the same mathematical procedure that was explained in Chap. 11 and which led to the conclusion that the mixture may fall apart into two phases. In the present case our mixed populations of hawks and doves may fall apart into colonies employing strategies A or B. The mathematics is entirely analogous to the one in mixtures and the results are the same ones, except for minor changes in notation. In particular the availability has the form

$$\mathfrak{A} = Nxg_A(\tau, p, z_A) + N(1-x)g_B(\tau, p, z_B) \,, \qquad (20.29)$$

where x is the fraction of the population living in colony A where strategy A is employed. The minimal availability is thus found by convexification of the graph $min[g_A, g_B]$ and the equilibrium values of x and $\mathfrak{A}$ result from

$$x^E = \frac{z - z_B^E}{z_A^E - z_B^E} \,, \qquad (20.30)$$

$$\frac{\mathfrak{A}_E}{N} = g_B(\tau, p, z_B^E) + \frac{z - z_B^E}{z_A^E - z_B^E} \left[g_A(\tau, p, z_A^E) - g_B(\tau, p, z_B^E) \right] \,, \qquad (20.31)$$

in complete analogy with (11.8), (11.9).

The free enthalpies $g_i = u_i - \tau s_i + p\frac{1}{n_i}$ as functions of τ, p and z may be composed from (20.13), (20.18) and (20.23), (20.24).

We obtain

$$g_A(\tau, p, z) = \tfrac{15}{2}\tau z + \tfrac{9}{2}\tau(1-z) - 7\tau z(1-z) -$$

$$-\tau\left\{ \left(s_A^H(\tau_R, p_R) + \left(\tfrac{15}{2} + \alpha_A^H\right)\ln\tfrac{\tau}{R} - \alpha_A^H \ln\tfrac{p}{p_R}\right)z + \right.$$

$$+ \left(s_A^D(\tau_R, p_R) + \left(\tfrac{9}{2} + \alpha_A^D\right)\ln\tfrac{\tau}{\tau_R} - \alpha_A^D \ln\tfrac{p}{p_R}\right)(1-z) -$$

$$\left. -\alpha_A^H z \ln\frac{\alpha_A^H z}{\alpha_A^H z + \alpha_A^D(1-z)} - \alpha_A^D(1-z)\ln\frac{\alpha_A^D(1-z)}{\alpha_A^H z + \alpha_A^D(1-z)} \right\}$$

$$+\alpha_A^H \tau z + \alpha_A^D \tau(1-z)$$

$$g_B(\tau, p, z) = \left(9\tau - \tfrac{3}{2}\tau^2\right)z + \left(\tfrac{51}{10}\tau - \tfrac{3}{5}\tau^2\right)(1-z) - \left(\tfrac{71}{10}\tau - \tfrac{101}{10}\tau^2\right)z(1-z) -$$

$$-\tau\left\{\left(s_B^H\left(\tau_R, p_R\right) + (9+\alpha_B^H)\ln\tfrac{\tau}{R} - 3(\tau - \tau_R) - \alpha_B^H\ln\tfrac{p}{p_R}\right)z + \right.$$

$$+ \left(s_B^H\left(\tau_R, p_R\right) + \left(\tfrac{51}{10} + \alpha_B^D\right)\ln\tfrac{\tau}{\tau_R} - \tfrac{6}{5}(\tau - \tau_R) - \alpha_B^D\ln\tfrac{p}{p_R}\right)(1-z) -$$

$$-\alpha_B^H z\ln\tfrac{\alpha_B^H z}{\alpha_B^H z + \alpha_B^D(1-z)} - \alpha_B^D(1-z)\ln\tfrac{\alpha_B^D(1-z)}{\alpha_B^H z + \alpha_B^D(1-z)}\left.\right\} +$$

$$+\alpha_B^H \tau z + \alpha_B^D \tau(1-z) \ .$$

$$(20.32)$$

20.7 Strategy equilibria in pure populations

If we write (20.32) for the pure dove population $z = 0$ and the pure hawk population $z = 1$ we obtain $p(\tau)$ curves for strategy equilibria in the pure species from

$$g_A\left(\tau, p, z = 0\right) = g_B(\tau, p, z = 0) \quad \text{and} \quad g_A(\tau, p, z = 1) = g_B(\tau, p, z = 1). \quad (20.33)$$

We introduce the pressure coefficients α from Table 20.1 and obtain with

$$(\tau_R, p_R) = (1, 1)$$

for doves: $p = \tau^{\frac{6}{5}}\exp\left[-\tfrac{4}{5} - \tfrac{1}{5}\tau - \tfrac{1}{3}\left[s_A^D(1,1) - s_B^D(1,1)\right]\right]$

$$(20.34)$$

for hawks: $p = \tau^{\frac{3}{2}}\exp\left[-\tfrac{1}{2} - \tfrac{1}{2}\tau - \tfrac{1}{3}\left[s_A^H(1,1) - s_B^H(1,1)\right]\right]$.

Without loss of generality we have chosen $(\tau_R, p_R) = (1, 1)$. In order to determine the $p(\tau)$-relations explicitly we need *one* observed (p, τ)-pair for each pure population, because that will provide us with specific values for the differences $[s_A^D(1,1) - s_B^D(1,1)]$ and $[s_A^H(1,1) - s_B^H(1,1)]$. We assume that we have observed the strategy transition in the dove population to occur at $(\tau, p) = (1, 1)$ and in the hawk population also at $(\tau, p) = (1, 1)$.[2] The s-differences are then equal to -3 in both equations (20.34)

$$s_B^D(1,1) = s_A^D(1,1) + 3 \quad \text{and} \quad s_B^H(1,1) = s_A^H(1,1) + 3,$$

and we may plot the transition curves $p(\tau)$. The result is shown in Fig. 20.2.

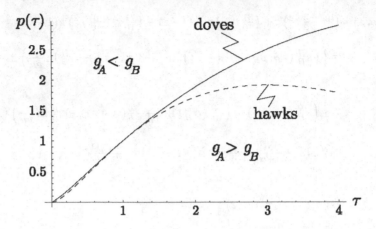

Fig. 20.2 Strategy transitions $p(\tau)$ for pure populations.
[The pair $(p, \tau) = (1, 1)$ represents the assumed
observational input, cf. text.]

The area above the curves of Fig. 20.2 represents states with $g_B > g_A$, so that strategy A prevails. We conclude that strategy A is the "low-price-strategy" while strategy B is the "high-price-strategy". This observation makes good sense also in an intuitive manner. Indeed, since the fighting of hawks and the posturing of doves in strategy A represent luxuries – as we have explained – these types of behaviour may be affordable in good times, when the price is low. In hard times, when the price is high, those luxuries need to be curtailed and strategy B is therefore the more appropriate strategy.

We remark that the graphs of Fig. 20.2 are analogues to the saturation vapour pressure curves of thermodynamics for pure constituents. In that case the areas below the curves correspond to the vapour phase and the areas above correspond to the liquid phase.

---

[2] Lacking the observation we make that assumption for simplicity. In reality it is quite unlikely that both the pure hawks and the pure doves should change their strategy at the same pair (p, τ).

20.8 Strategy diagrams

We insert the pressure coefficients α according to Table 20.1 into the free enthalpies g_A, g_B in (20.32). Also in those equations we insert $(\tau_R, p_R) = (1, 1)$ and we replace $s_A^D(1,1)$ and $s_A^H(1,1)$ by use of (20.35). That leaves us with an unknown function

$$\tau \left[s_B^D(1,1)(1-z) + s_B^H(1,1)z \right]$$

in both free enthalpies g_A and g_B. Since we are only interested in the free enthalpy differences, we ignore that function.

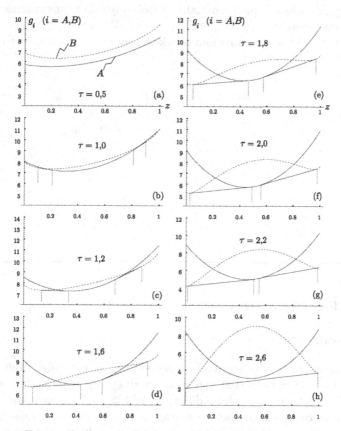

Fig. 20.3 Free enthalpies for strategy A (solid) and strategy B (dashed) for eight different prices τ as functions of the hawk fraction z.

We fix p to be given by the reference value $p_R = 1$ and obtain

$$g_A(\tau, z) = \left(8 - 4z + 7z^2\right)\tau -$$
$$-\tau\left[(5 + 3z)\ln\tau + 0.5\left(z\ln z + (1-z)\ln(1-z)\right)\right]$$
$$g_B(\tau, z) = \left(7.4 - 5z + 7.1z^2\right)\tau + \left(0.6 + 11z - 10.1z^2\right)\tau^2 -$$
$$-\tau\left[(8.6 - 3.9z)\ln\tau + 3.5\left(z\ln z + (1-z)\ln(1-z)\right)\right] .$$

These functions are plotted in Fig. 20.3 for eight different values of τ. Inspection shows that the low price strategy is strategy A, irrespective of the hawk fraction. For intermediate and high values of τ the graphs of g_A and g_B intersect which provides the possibility of convexification. We have explained before how this works through the construction of common tangents which are indicated in the figure. The projection of the tangents into the (τ, z)-strategy-diagram has also been explained before. The strategy diagram relevant to the present case is shown in Fig. 20.4. The diagram must be compared with the one in the lower right corner of Fig. 20.1. There are some obvious differences which are all due to the fact that we have now taken care of the entropy. The entropy opposes complete unmixing absolutely and that explains that we have two new areas in Fig. 20.4, viz. the areas V and VI in which we find homogeneously mixed dove-rich and hawk-rich populations respectively.

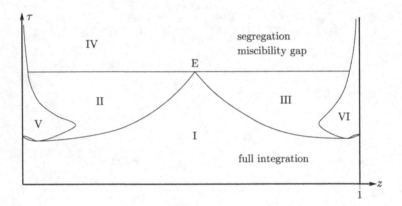

Fig. 20.4 Strategy diagram with the effects of entropy

20.9 Discussion

The entropic opposition to unmixing is something we recognize from thermo-dynamics. But other than that the strategy diagrams of Figs. 20.1 and 20.3 are quite different from the phase diagrams of Fig. 11.1. Indeed, the miscibil-ity gap is now on *top* while before it was at the *bottom*. The reason for the difference is not entropic but rather energetic – or due to the shortfall. Let us consider:

In thermodynamics it is always true that high temperature makes the entropy dominate over the energy. Energy loses the influence, because the atoms or molecules may jump out of their potential wells. We have exhibited many examples for that phenomenon.

In our sociological model, however, the energy – or rather the shortfall – dominates for high prices, and the entropy can achieve little more than prevent complete unmixing. Our rules of the game make the entropy pale into insignificance compared to the shortfall when the price grows big.

One peculiarity of the phase diagram of Fig. 20.4 should be mentioned, although we shall not discuss it in detail. The inverted "peaked caps" of areas II and III contain lateral loops. This is due to the situation shown in Fig. 20.3b, where the curves A and B intersect *and* have the same end-points. A situation like that occurs quite frequently in thermodynamics − although not in this book. The phenomenon is known as *azeotropy* and the point where the osculating curves touch each other is called the azeotropic point. The present diagram has two such points.

21

History of thermodynamics

21.1 Temperature

Not surprisingly medical doctors were the first people interested in temperature, because the body temperature of a patient provided some sign of his well-being, or rather ill-being. Of course some standard of normal body temperature was needed and in this respect the doctor Johannis Hasler from Berne had a quaint observation to offer. In his book "De Logistica Medica" in 1578 he published a table of body temperatures of people living at different latitudes. He attributed the value 9 to dwellers of the tropics, 5 to us in Europe and 1 to the eskimos, cf. Fig. 21.1.

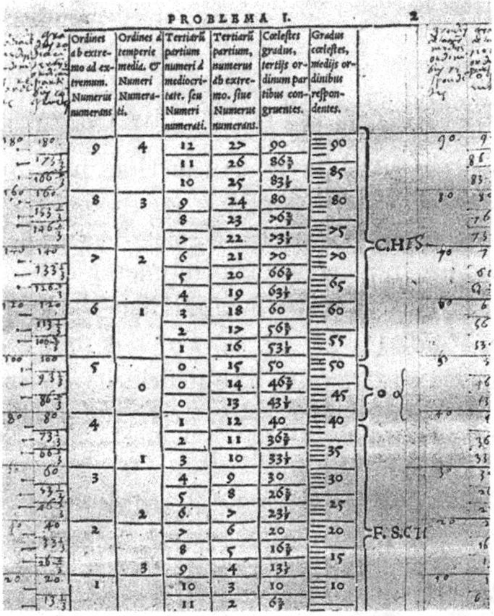
PROBLEMA I. 2

Ordines ab extremo ad extremum. Numerus numerans	Ordines à temperie media. & Numeri Numerati.	Tertiarũ partium numeri à mediocritate. seu Numeri numerati.	Tertiarũ partium, numerus ab extremo. siue Numerus numerans.	Cœlestes gradus, tertijs ordinum partibus congruentes.	Gradus cœlestes, medijs ordinibus respondentes.
9	4	12	27	90	90
		11	26	86 2/3	
		10	25	83 1/3	85
8	3	9	24	80	80
		8	23	76 2/3	
		7	22	73 1/3	75
7	2	6	21	70	70
		5	20	66 2/3	
		4	19	63 1/3	65
6	1	3	18	60	60
		2	17	56 2/3	
		1	16	53 1/3	55
5		0	15	50	50
	0	0	14	46 2/3	
		0	13	43 1/3	45
4		1	12	40	40
		2	11	36 2/3	
	1	3	10	33 1/3	35
3		4	9	30	30
		5	8	26 2/3	
	2	6	7	23 1/3	25
2		7	6	20	20
		8	5	16 2/3	
	3	9	4	13 1/3	15
1		10	3	10	10
		11	2	6 2/3	

Fig. 21.1 Hasler's table of body temperatures

Misconceptions like that about temperature were rife at that time but they were slowly being eliminated. Thus the Venetian diplomat Gianfrancesco Sagredo used a thermometer to measure the temperature of well-water over the year. He was thus able to overcome one prejudice and he put the result of his observations into a letter to his teacher and mentor, Galileo Galilei (1564 - 1642), cf. Fig. 21.2.

GIOVANFRANCESCO
SAGREDO a GALILEO
in Firenze.
Venezia, 7 febbraio 1615

Molto Ill.^{re} Sr Ecc.^{mo}
...Con questi istrumenti ho chiaramente veduto, esser molto più freda l'aqua de' nostri pozzi il verno che l'estate; e per me credo che l'istesso avenga delle fontane vive et luochi soteranei, ancorchè il senso nostro guidichi diversamente.
Et per fine li baccio la mano.

In Venetia, a 7 Febraro 1615.
Di V.S. Ecc.^{ma}

Tutto suo Il Sag.
Mi perdoni: non ho tempo di riveder queste.

Fig. 21.2 Galileo Galilei. A cut from a letter of Sagredo to Galilei with the remarkable sentence:"...I have clearly seen that well-water is colder in winter than in summer..., although our senses tell differently."

One trouble was that no decent thermometer existed in the early days. To be sure some instruments were around like the air thermometer of the Welsh doctor Robert Fludd (1574 - 1637) which is shown in Fig. 21.3$_{left}$. Fludd refused to be drawn into the dispute about priority of the invention by saying

"...the instrument has many counterfeit masters or patrons in this our age, who because they have a little altered the shape of the modell, do vainly glory and give out, that it is a masterpiece of their own finding out."

And he continues by stating that he has read about a thermometer

"in a manuscript of five hundred years of antiquity at the least."

Inspection of Fig. 21.3₍ₗₑ𝒻ₜ₎ shows that the reading of Fludd's thermometer could be biased by air pressure. Such a bias was prevented in sealed thermometers like the Florentine thermometer shown in Fig. 21.3$_{right}$.

Fig. 21.3 Left: Fludd's open thermometer
Right: Florentine sealed thermometer.

Since no two thermometers were exactly alike in those early days it was difficult to communicate quantitative observations from one place to another. What was needed were fix points, so that the instruments could be gauged. Some interesting propositions for fix points were these:

> Freezing point of water and melting point of butter,
> temperature of a salt-ice mixture and of a deep cellar,
> melting point of ice and human body temperature.

The latter is due to Isaac Newton (1642 - 1727) who apparently thought little of Hasler's table, if indeed he knew it. Newton divided the interval between his fix points into 12 equal parts and placed the melting point at 0 and body temperature at 12.

The next step came with the tradesman Daniel Gabriel Fahrenheit (1642 - 1727) from Danzig. He had developed the notion that three fix points should be better than two. And he chose

> the temperature of a mixture of ice, water and "seasalt",
> the temperature of such a mixture but without the salt,
> the temperature in the "armpit of a living man in good health".

Those temperatures he calls 0°F, 32°F and 96°F. Asimov speculates that the number 96 reflects Newton's division into 12 parts of which each one was split into 8 steps by Fahrenheit. Later Fahrenheit adjusted the scale slightly so as to put the boiling point of water to 212° − exactly 180° above the freezing point without seasalt. In this manner the human body temperature ended up at 98.6°F. The Fahrenheit scale is still used in the United States of America.

In most of the rest of the world the customary scale goes back to the Swedish astronomer Aulus Cornelius Celsius (1701 - 1744) who fixed his 0°C to the boiling point of water and 100°C to the melting point of ice. Thus Celsius counted "backwards". His successor in the observatory of Uppsala was Mårten Strömer (1707 - 1770) and it was he who changed the direction on April 13, 1750, six years after Celsius' death. That day marks the beginning of what we now call the "Celsius" scale, or centigrade scale.

From the data given above it is easy to calculate back and forth between the Celsius scale and the Fahrenheit scale by the formula $C = \frac{5}{9}(F-32)$.

21.2 The first law

The American man-of-the-world and globetrotter Benjamin Thompson (1753 - 1814), later Graf von Rumford − ennobled by Karl Theodor, elector of Bavaria − was first to cast doubt upon the caloric theory. That theory goes back to the great chemist Antoine Laurent Lavoisier (1743 - 1794) and it asserts that heat is a weightless fluid, the *caloric*. The heating observed by lifting chips from a metal on a lathe (say) was considered as caloric liberated from the material.

Posthumously Rumford crossed Lavoisiers' path. He went to Paris and married Lavoisier's widow. Asimov[1] writes:"The marriage was unhappy. After four years they sepa-rated and Rumford was so ungallant as to hint that she was so hard to get along with that Lavoisier was lucky to have been guillotined.[2] However it is quite obvious that Rumford was no daisy himself."

Fig. 21.4 Graf von Rumford

---

[1] Asimov's Biographical Encyclopedia of Science and Technology. Pan Reference Books, London & Sydney (1978).
Americans do not like their countryman Graf Rumford, because in the war of independence he fought on the British side.

[2] That lady was also the recipient of Lavoisier's last letter, written on the eve of his execution. The chemist was being philosophical. He writes: "It is to be expected that the events in which I am involved, will spare me the inconvenience of old age."

Rumford arranged the English garden in Munich for the elector, he fed the Bavarian poor by the filling "Rumford soup" and he had cannon barrels bored. During this latter activity he observed that a dull drill set more heat free than a sharp one although no chips appeared. He says: "It is difficult for me, if not downright impossible, to think anything else than that heat is motion, because that is what was supplied to the barrel as the heat appeared in it".

Maybe this is not too clear but it becomes clear that Rumford was on the right track when we savour his attempts to formulate the *mechanical equivalent of heat*. He observes that the drill was kept in rotation by two horses pulling on a capstan-bar and the heating of the barrel was "equivalent at least to the heating of two big wax candles". He thinks little, however, of the efficiency of the conversion, because "one should have obtained more heat by burning the fodder of the horses".

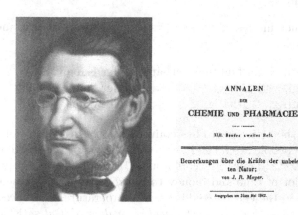

Fig. 21.5 Robert Julius Mayer. Cut from the title page
of his first published paper.

Robert Julius Mayer (1814 - 1878) was the first who had a clear idea that heat was a form of energy and that energy is conserved but may be converted from heat to mechanical energy. In the manner of the 19th century he formulates that knowledge succinctly by saying

Ex nihilo nil fit, nil fit ad nihilum.

Fig. 21.5 shows a portrait of Mayer next to a cut from the title page of his first published work. That paper contains the sentence: "... and it follows, that the drop of a weight from a height of 367 m corresponds to a heating of the same weight of water from 0° to 1°." Later, 1845 Mayer formulated that observation in shortened form thus:

$$1° \text{ heat} = 1 \text{ gram at } \begin{cases} 367 \text{ m} \\ 1130 \text{ Parisian foot} \end{cases} \text{ height.}$$

This is not too bad as the mechanical equivalent of heat. Nowadays the older ones among the readers would express this by saying[3]

$$1 \text{ cal} = 4.18 \text{ J.}$$

As a medical doctor Mayer did not share the sober jargon of the professional scientists. Therefore his works are full of the most colorful hyperbolic statements. Mayer is first to think that the sun's heat energy stems from the potential or kinetic energy of the masses falling into it. In his words this sounds as follow.

> "...these masses tumble into their joint tomb with a mighty crash. And since no cause exists without an effect, each one of the cosmic masses must produce a certain positive heat." "...the fruit of this is the most marvellous phenomenon of the material world, the eternal source of light."

Mayer was much interested in the physiological effects of heat and he cites body temperatures

bei einem Neger faul und unthätig in der Kabane 37°
    desgl.   desgl.   in der Sonne   40.20°
    desgl.   thätig   in der Sonne   39.75° "

The idea is that the active negro has converted some heat into work.[4]

Usually three people are named as the discoverers of the first law. Mayer, Joule and Helmholtz. James Prescott Joule (1818 - 1899) was privately rich and he spend a lot of time and money to improve the measurement of temperature so that in the end he could detect temperature differences down to 0.005°F. Also he measured the *mechanical equivalent of heat* satisfactorily, so that he gets the credit and the unit of energy is named after him.

---

[3] Of course the calory is now an obsolete unit. Formerly it defined the heat needed to raise the temperature of 1 gram of water by 1 degree Celsius. It was a relief to everybody when the caloric units disappeared altogether — in 1972 — and energy of whatever form was measured in Joule.

[4] Mayers student days coincided with a repressive political period in Germany. He did not toe the line and was forced to flee the country. Therefore he took hire as a doctor on a ship to Java. That is how he met negroes in and out of their cabin.

Joule was almost a fanatic on the subject of measurement and even on his honeymoon he took time out to devise a special thermometer to measure the temperature of the water at the top and bottom of a scenic waterfall his wife and he were to visit. The idea was that the energy of falling water should be converted to heat once it was stopped.[5]

Fig. 21.6 James Prescott Joule

Although Joule recognized the principle of conservation of energy, and so did Mayer before him, the first to present it to the world as an explicit generalization was Hermann Ludwig Ferdinand von Helmholtz (1821 - 1894). Therefore it is usually Helmholtz who gets the credit.[6]

Helmholtz was yet another physician scientist. He studied the working of the eye and the ear and formulated the "Helmholtz vortex theorems", mathematically non-trivial results for his time. Lenard[7] says: "...that Helmholtz, who had no formal mathematical education, was able to do this, shows the absolute uselessness of the extensive mathematical instruction in our universities, where the students are tortured with the most outlandish ideas, ... when only a few are capable of getting results with mathematics, and those do not even need this endless torment."

Fig. 21.7 Hermann Ludwig Ferdinand von Helmholtz

---

[5] Asimov, loc. cit.

[6] Asimov. loc. cit.

[7] P. Lenard. Große Naturforscher. J.F. Lehmann Verlag Münschen 1941. Students of thermodynamics usually like this remark.

Helmholtz also made Mayer's ideas about the source of solar heat more concrete. He assumed that the sun's radius contracts and calculated the consequent heating. Assuming that the sun radiates at a constant rate, he found that its radius would have been equal to the radius of the earth's orbit 25 million years ago. Therefore the earth had to be younger than that, which was incredible to the geologists and biologists. Helmholtz' calculation was faultless but his and Mayer's premise was wrong. Both could not have known that the source of the solar energy was nuclear.

It is easy to argue that the law of conservation of energy is the most important law of physics. So, how was its discovery received?

Mayer sent his first paper, written in 1841, to "Poggendorff's Annalen" and recieved no answer despite several reminders. The paper was never published. It was found after Poggendorff's death in his desk. Mayer was deeply disappointed and wrote a new paper, cf. Fig. 21.5, which he sent to Justus von Liebig, editor of the "Annalen der Chemie und Pharmacie". We have reproduced his covering letter in Fig. 21.8 because it might teach the reader − at least those of them who have some command of German − how editors should be approached. The paper was accepted.

Hochverehrtester Herr Professor!
Beifolgenden kurzen Aufsatz bin ich so frei, Euer Wohlgeboren zur gefälligen Aufnahme in Ihren Annalen für Chemie und Pharmazie vorzulegen. − − −

− − − −, Möchte es Euer Wohlgeboren gefallen, sich vom Durchlesen des kleinen Aufsatzes durch den vielleicht barock scheinenden Anfang nicht abhalten zu lassen und möchten Sie solchen der Erhaltung eines Plätzchens in Ihren weltverbreiteten Annalen für würdig erachten. Sollten Sie indessen von demselben keinen Gebrauch zu machen geneigt sein, so erlaube ich mir die Bitte um gefällige Zurücksendung des Manuskriptchens.

Mit vorzüglichster Hochachtung
verharre ich
Euer Wohlgeboren gehorsamster Diener
Dr. J. R. Mayer

Fig. 21.8 Cut from a letter of Mayer to Liebig on March 31, 1842

Things went worse with Joule. Asimov writes[8]: "The statement of his discovery was rejected by various learned journals as well as by the Royal Society and he was forced to present it in a public lecture in Manchester and then get his speech published in full by a reluctant Manchester newspaper on which his brother was music critic."

Helmholtz fared no better. He had difficulties to publish. Once again it was Poggendorff who disliked it and disposed of it as "mere philosophy".[9] In

---

[8] Asimov. loc. cit.

[9] Dokumente der Wissenschaftsgeschichte 1. C. Kirsten, K. Zeisler (eds). Akademie Verlag, Berlin (1982).

the end Helmholtz published his work as a pamphlet, cf. Fig. 21.9. The title
"On the conservation of force" illustrates neatly that the concept of energy
had not fully jelled in the minds of the scientists of the time.

To be sure, all three discoverers of the 1st law lived long enough to recieve
recognition eventually. So also Mayer:

> "In the year 1867 his king, whose pleasure it is to reward merits,
> created him a Knight of the Order of the Crown of Württemberg.
> And in 1869 he was made an honorary member of the Chamber of
> Commerce of his home town Heilbronn." [10]

Mayer could now call himself "von Mayer".

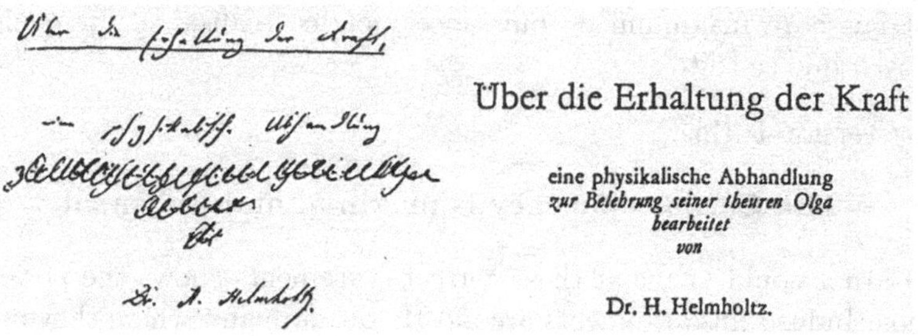

Über die Erhaltung der Kraft

eine physikalische Abhandlung
zur Belehrung seiner theuren Olga
bearbeitet
von

Dr. H. Helmholtz.

Fig. 21.9 Title page of Helmholtz' pamphlet.
[The dedication to "dear Olga" was
scratched out before printing.]

21.3 The second law

Apart from Clausius whose ideas we have discussed in detail, one must mention the French physicist Nicolas Léonard Sadi Carnot (1796 - 1832) as a forerunner of the second law. Carnot's great work "Reflexions sur la Puissance
Motrice du Feu et sur les Machines propres à développer cette Puissance" appeared in 1824. At that time thousands of steam engines were already working
and their efficiency was less than 10%.

Therefore Carnot asked whether and how the efficiency might be improved.
Everything appeared possible: One might assume that some process other
than the cycle with two adiabates and two isobars might be more efficient or,
perhaps, that the use of some substance other than water might be beneficial.
In this matter Carnot came to two conclusions.

- The biggest possible efficiency of a heat engine, whose working substance
  assumes temperatures between T_L and T_H in the process, is the engine

---

[10] H. Schulz, H. Weckbach. Robert Mayer. Sein Leben und sein Werk. Publication
of the Archives of the City Heilbronn (1964).

which we now call the Carnot engine, which exchanges heat *only* at those temperatures. Carnot says that that process is "le plus avantageux possible, car il ne s'est fait aucun rétablissement inutile d'équilibre dans le calorique."

- The efficiency of the Carnot engine is given by

$$e = 1 - \frac{T_L}{T_H}$$

It is independent of the working substance and of the pressure- and volume-range covered by the process. In Carnot's words

"Le maximum de puissance motrice résultant de l'emploi de la vapeur est aussis le maximum de puissance motrice réalisable par quelque moyen que ce soit."

A short version is this:

The Carnot efficiency is maximal and universal.

That Carnot could arrive at these correct statements shows the power of his intuition. Indeed, his arguments are mostly obscure and where they are clear, they are wrong, because Carnot believed in the caloric theory, cf. Fig. 21.10.

Nous suppeserons ... que les quantités de chaleur absorbées et dégagées dans ce diverses transformations sont exactement compensées. Ce fait n'a jamais été révoqué en doute; il a été d'abord admis sous reflexions et vérifié ensuite dans beaucoup de cas par les experiences du calorimètre. Le nier ce serait renverser toute la thèorie de la chaleur, à quelle il sert de base.

Dans la chute d'eau, la puissance motrice est rigoureusement proportionelle à la différence de niveau entre le réservoir supérieur et le réservoir intérieur. Dans la chute du calorique, la puissance motrice augmente sans doute avec la différence de température entre le corps chaud et le corps froid.

Fig. 21.10 Sadi Carnot and the caloric theory

Clausius keeps his criticism mild when he talks about Carnot's ideas about the conversion of heat to work. He understood their deficiencies but he accepts the salient role of Carnot engines. Then he proceeds to put things right as we have explained in Chap.1.

Let us, however, mention some of the reactions to Clausius' concepts and, in particular, to his extrapolation about the heat death. The reactions were not all positive.

Johannes Joseph Loschmidt (1821 - 1985), an Austrian physicist deplores

> "the terroristic nimbus of the second law .... which lets it appear as a destructive principle of all life in the universe."

The American historian Henry Adams (1838 - 1918) cites Clausius and summarizes his view by saying

> "for the layman this only means that the heap of ashes becomes ever bigger."

Adams was well-known for his pessimistic views; he was a person who considered optimism as a symptom of idiocy.

It is often said that the world goes in a circle ... such that the same states are always reproduced. Therefore the world could exist forever. The second law contradicts this idea most resolutely ... The entropy tends to a maximum. The more closely that maximum is approached, the less cause for change exists. And when the maximum is reached, no further changes can occur; the world is then in a dead stagnant state.

Fig. 21.11 Rudolf Clausius and his contemplation of the heat death

Oswald Spengler (1880 - 1936), the philosopher historian and author of the work "The Decline of the West" devotes a whole chapter to entropy and says this:

> "The end of the world as the completion of an inevitable evolution — that is the twilight of the gods. Thus the doctrine of entropy is the last, irreligious version of the myth."

Well, let us not go into that too deeply. The fact is though that the doctrines of conservation of energy and growth of entropy have penetrated the western conciousness, even though they might often be expressed rather flippantly in a free and easy manner thus:

1st law: You can't win
2nd law: You can't even break even.

21.4 The statistical interpretation of entropy as $S=k\ln W$

Whatever criticism Clausius had to suffer for having conceived entropy it was like nothing compared to what befell Boltzmann when he provided a plausible interpretation for entropy and its growth within his kinetic theory of gases. We have laid out that interpretation in some detail in previous chapters, cf. Chaps. 3 and 16. It is the only way to understand entropy in a suggestive manner.

To go straight to the deepest depth I went for Hegel; what unclear thoughtless flow of words I was to find there! My unlucky star led me from Hegel to Schopenhauer. Even in Kant there were many things that I could grasp so little that given his general acuity of mind I almost suspected that he was pulling the readers's leg or was even an impostor.[11]

Fig. 21.12 Ludwig Edward Boltzmann. His tombstone in Vienna's central cemetary.
Also: Boltzmann's opinion about philosophers.

Boltzmann had to fight on two fronts because there were two major objections to the growth of entropy, viz.

the **reversibility objection** and the **recurrence objection**.

Atomic and molecular processes are reversible, i.e. they can proceed forward or backward along the same orbit, if only initial conditions are inverted.

---

[11] Flamm. Stud. Hist. Phil. Sci.
14, p. 257 (1983).

And therefore, when the entropy grows in the forward process it must decrease in the backward process. From this Loschmidt concluded that the entropy grows **just as often** as it decreases and consequently there could not be any macroscopic growth of entropy. Loschmidt was the main proponent of the *reversibility objection.*

Boltzmann replied, that there are many more disordered states than ordered states, where the occurrence of each state is equally probable. Therefore, if we start in an ordered state, the entropy-increasing process order → disorder will nearly always be seen. On the other hand, if we start in a disordered state, the entropy-decreasing process disorder → order will *almost never* be seen. Therefore entropy-increasing processes occur **more often** than entropy-decreasing ones.

1. *Die Energie des Weltalls ist konstant.*
2. *Die Entropie des Weltalls strebt einem Maximum zu.*

Fig. 21.13 Josiah Willard Gibbs. A quote from Clausius
which Gibbs placed in front of his great work
"On the equilibrium of heterogeneous substances".

Notabene: *nearly* always and *almost* never. Indeed, the second law is not a deterministic law for single atoms; rather it is a statistical law for many atoms which is obeyed with great probability. The American physicist Josiah Willard Gibbs (1839 - 1903) says:

"The impossibility of an uncompensated decrease
of entropy seems to be reduced to an improbability."

The recurrence objection was different. It was based on a result by the French mathematician Jules Henri Poincaré (1854 - 1912). Once a mechanical system is set in motion, it will in time return to a state that is arbitrarily close to the initial state. This is obviously in contradiction to the monotone growth of entropy to an equilibrium.

Once again the resolution of the contradiction lies in the statistical character of the entropy. Poincaré's theorem is correct and the entropy principle is *nearly* correct. We have discussed this situation in Chap. 5 in connection with the rubber molecule: Since each state has the same − very small − probability of occurring, it will occur over and over again, if only after long long time intervals. This is also true for the initial state, even though it may have a smaller entropy than the intermediate states. However the recurrence times which we can calculate for simple models, are many orders of magnitude bigger than the likely age of the universe. So nobody is worried, because nobody dares to think that far ahead.

The statistical character of thermodynamics was difficult to grasp for some people. One who refused even to try was Ernst Zermelo who involved Boltzmann in an acrimonious public debate. We shall not follow this struggle except to quote a sentence from one of Boltzmann's rebuttals, or − in Zermelo's words − "so-called" rebuttals. That sentence is interesting because it reveals what Boltzmann thought about time. He speculates that the universe is in thermal equilibrium but subject to fluctuations which may last the relatively short time of "a few eons". And then he continues: "There is no difference for the universe as a whole between "forward" and "backward" directions of time. But for those worlds on which life exists − and which are therefore in a relatively improbable state − the direction of time is determined by the direction of growing entropy, which points from less probable states to more probable ones." [12]

James Clerk Maxwell (1831 - 1979) had no difficulty to understand the probabilistic nature of the entropy growth. He invented the Maxwell demon to perhaps outwit chance.

> "... a creature whose talents are so extraordinary, that it can trace the orbit of each molecule. The demon guards a door between two volumes of a gas with different temperatures. It allows only the fast molecules from the cold side to pass and only the slow molecules from the hot side."

Obviously in that manner − if such a demon existed − the warm side becomes warmer and the cold side colder and entropy would thus decrease.

Maxwell and Boltzmann corresponded but there seems to have been a slight feeling of disdain between them, cf. Fig. 21.14. This is not unusual in this day and age either between scientists working in the same field − even on a lesser level.

---

[12] L. Boltzmann: Zu Hrn. Zermelos Abhandlung "Über die mechanische Erklärung irreversibler Vorgänge." Ann. Phys. **59** Leipzig (1896).

Boltzmann about Maxwell:...
immer höher wogt das Chaos der
Formeln.

Maxwell about Boltzmann: ...
I am much inclined to put the
whole business in about six lines.

Fig. 21.14 James Clerk Maxwell

The statistical interpretation of entropy has provoked scientists outside the fields of physics and chemistry to extrapolate the notion to their fields. Economists use it to characterize the distribution of money and goods and ecologists interpret the dissipation of natural resources as entropy production. And an original wit has illustrated the progress of western culture to more and more disorder by the graffitto

> Hamlet: to be or not to be
> Camus: to be is to do
> Sartre: to do is to be
> Sinatra: do be do be do be do.

21.5 Third law

Thermodynamics is the only classical theory in which the Planck constant plays a significant role. That constant determines the additive entropy constants in a body. The realization of that fact came in stages and Nernst's law, or Nernst's theorem − the third law of thermodynamics − represented the first stage. The law states that the entropies of all bodies become independent of pressure and of the prevailing phase when the absolute zero of temperature is approached.

Modesty was not Nernst's forte, cf. Fig. 21.15. Asimov[13] describes him as a kind of confidence man:

> "His most famous invention was a lamp made of ceramics ... However
> it had disadvantages and was no competitor to Edinson's light. And
> yet, to Edinson's astonishment he sold the patent for a million marks."

[13] Asimov, loc. cit.

Actually the third law had also two discoverers, because Planck strengthened the law considerably by recognizing that the entropies of all bodies are zero at absolute zero.

Nernst reassures us concerning the emergence of further thermodynamic laws:

The 1ˢᵗ law had three discovers: Mayer, Joule and Helmholtz.

The 2ⁿᵈ law had two discovers: Carnot and Clausius.

The 3ʳᵈ has only one discoverer, namely himself: Nernst.

The 4ᵗʰ law ... (?)

Fig. 21.15 Hermann Walter Nernst (1864 - 1941)

A new scientific truth does not triumph by convincing its opponents and making them see the light, but rather because its opponents finally die, and a new generation grows up that is familiar with it.

Fig. 21.16 Max Karl Ernst Ludwig Planck

This was not Planck's greatest contribution to thermodynamics, of course. His greatest contribution was that he firmly incorporated the theory of radiation into thermodynamics. And in order to do that he had to conceive of energy quanta. Thus be became one of the founders of quantum mechanics and modern physics. In due course the Planck constant was then recognized as a determining factor in the entropy constant.

Appendix. Equation of balance

22.1 Transport theorem

Let $V_\mathbf{v}$ and $V_\mathbf{u}$ be two volumes which coincide at time t but which are different in their further development: The surface of $V_\mathbf{v}$ moves with the velocity $\mathbf{v}$ of a body, while the surface of $V_\mathbf{u}$ moves with the arbitrary velocity $\mathbf{u}$. Therefore at time t the rate of change of a generic quantity

$$\Psi = \int_V \rho\psi dV \qquad (A.1)$$

inside $V_\mathbf{v}$ or $V_\mathbf{u}$ — momentarily equal — are given by

$$\frac{\mathrm{d}}{\mathrm{d}t} \int_{V_\mathbf{v}} \rho\psi dV = \int_V \frac{\partial \rho\psi}{\partial t} dV + \int_{\partial V} \rho\psi v_i n_i dA$$

$$\frac{\mathrm{d}}{\mathrm{d}t} \int_{V_\mathbf{u}} \rho\psi dV = \int_V \frac{\partial \rho\psi}{\partial t} dV + \int_{\partial V} \rho\psi u_i n_i dA$$

$$(A.2)$$

V and ∂V in (A.1), (A.2) are the equal values of $V_\mathbf{v}$, $V_\mathbf{u}$ and $\partial V_\mathbf{v}$, $\partial V_\mathbf{u}$ at time t. $\mathbf{n}$ is the outer unit normal. The equations (A.2) are known as *Reynolds' transport theorem*. Figure A.1 is an illustration of that theorem.

The surface $\partial V_\mathbf{v}$ is said to confine a closed system, because it always contains the same mass, while the surface $\partial V_\mathbf{u}$ contains an open system, since generally mass passes through the surface.

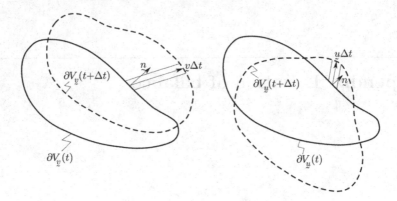

Fig. A.1 Visualization of the transport theorem.
The volumes $V_{\mathbf{v}}$ and $V_{\mathbf{u}}$ coincide at time t
but they move differently.

The transport theorem is less obvious in the presence of a singular surface s
that cuts the closed volume $V_{\mathbf{v}}$ into two open volumes V_- and V_+, cf. Fig. A.2.
That singular surface may be an impermeable wall, or a permeable membrane,
or a phase boundary. Its unit normal vector is $\mathbf{e}$, its velocity $\mathbf{u}$ and the fields
ρ and ψ have the values $\rho^\pm$, $\psi^\pm$ on its frontside and backside respectively.
Application of $(\text{A.2})_2$ to V_- and V_+ gives

$$\frac{\mathrm{d}}{\mathrm{d}t}\int_{V_-}\rho\psi\,\mathrm{d}V = \int_{V_-}\frac{\partial\rho\psi}{\partial t}\mathrm{d}V + \int_{A_-}\rho\psi v_i n_i \mathrm{d}A + \int_s \rho\psi u_i e_i \mathrm{d}A$$

$$\frac{\mathrm{d}}{\mathrm{d}t}\int_{V_+}\rho\psi\,\mathrm{d}V = \int_{V_+}\frac{\partial\rho\psi}{\partial t}\mathrm{d}V + \int_{A_+}\rho\psi v_i n_i \mathrm{d}A - \int_s \rho\psi u_i e_i \mathrm{d}A \ ,$$

where $A_\pm$ are the "caps" on the two sides of s which form part of ∂V, cf.
Fig. A.2. Summation gives

$$\frac{\mathrm{d}}{\mathrm{d}t}\int_{V_-+V_+}\rho\psi\,\mathrm{d}V = \int_{V_-+V_+}\frac{\partial\rho\psi}{\partial t}\mathrm{d}V + \int_{A_-+A_+}\rho\psi v_i n_i \mathrm{d}A - \int_s [\rho\psi]\,u_i e_i \mathrm{d}A \ , \quad \text{(A.3)}$$

where $[\rho\psi]$ stands for $\rho^+\psi^+ - \rho^-\psi^-$.

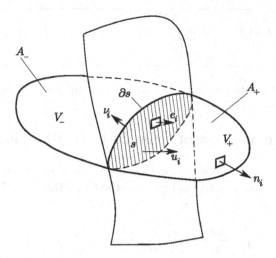

Fig. A.2 A closed volume V cut by a singular surface s into parts $V_\pm$.

22.2 Integral equations of balance

Ψ may change in time due to a flux of Ψ through the surface ∂V and due to a source of Ψ inside V.

For $V_\mathbf{v}$, the volume moving with the velocity $\mathbf{v}$, the rate of change of Ψ may be expressed by the generic equation of balance for a *closed system*, viz.

$$\frac{\mathrm{d}}{\mathrm{d}t} \int_{V_\mathbf{v}} \rho\psi\mathrm{d}V = - \int_{\partial V} \Phi_i n_i \mathrm{d}A + \int_V \sigma \mathrm{d}V \,,$$

$$(A.4)$$

where Φ_i is the non-convective flux density vector[1] of Ψ and σ is the source density.

Elimination of $\int_V \frac{\partial \rho\psi}{\partial t} \mathrm{d}V$ and $\frac{\mathrm{d}}{\mathrm{d}t} \int_{V_\mathbf{v}} \rho\psi\mathrm{d}V$ between (A.2) and (A.4) gives the generic equation of balance for an *open system*

$$\frac{\mathrm{d}}{\mathrm{d}t} \int_{V_\mathbf{u}} \rho\psi\mathrm{d}V = - \int_{\partial V} \left(\rho\psi\left(v_i - u_i\right) + \Phi_i\right) n_i \mathrm{d}A + \int_V \sigma \mathrm{d}V \,.$$

$$(A.5)$$

If the *system is open and at rest*, meaning that its surface ∂V is at rest, so that $u_i = 0$, equation (A.5) reads

---

[1] Φ_i is often briefly called the non-convective flux or just the flux, when no misunderstanding can arise.

$$\int_V \frac{\partial \rho\psi}{\partial t} dV = - \int_{\partial V} (\rho\psi v_i + \Phi_i) n_i dA + \int_V \sigma dV \,.$$

$$(A.6)$$

since the time derivative may be pulled into the integral in this case.

22.3 Local equations of balance in regular points

Given the proper smoothness properties, the surface integral in (A.6) may be converted into a volume integral by use of the Gauß theorem. Therefore (A.6) may be written as

$$\int_V \left(\frac{\partial \rho\psi}{\partial t} + \frac{\partial (\rho\psi v_i + \Phi_i)}{\partial x_i} - \sigma \right) dV = 0 \qquad (A.7)$$

and since this equation must hold for all volumes, even infinitesimally small ones, the integrand itself must vanish. Thus we obtain the generic *local equation of balance*

$$\frac{\partial \rho\psi}{\partial t} + \frac{\partial (\rho\psi v_i + \Phi_i)}{\partial x_i} - \sigma = 0 \,.$$

$$(A.8)$$

The prototype of that equation is the mass balance which results from setting $\Psi = 1$, $\Phi_i = 0$ and $\sigma = 0$ so that we obtain

$$\frac{\partial \rho}{\partial t} + \frac{\partial \rho v_i}{\partial x_i} = 0,$$

which is also known as the continuity equation, especially in fluid mechanics. It may be used to simplify the generic local balance equation (A.8) to read

$$\rho \frac{d\psi}{dt} + \frac{\partial \Phi_i}{\partial x_i} = \sigma, \qquad \text{where } \frac{d}{dt} = \frac{\partial}{\partial t} + v_i \frac{\partial}{\partial x_i} \qquad (A.9)$$

is the *material,* or *substantial derivative.*

22.4 Jump conditions

When there is a singular surface, it may have properties of its own; there may be surface distributions $\rho_s \psi_s$ and s_s of Ψ and its source, and there may be a line distribution Φ_i^s of the flux, such that we have

$$\int\limits_{V} \rho\psi dV \;=\; \int\limits_{V_-+V_+} \rho\psi dV \;+\; \int\limits_{s} \rho_s\psi_s dA$$

$$\int\limits_{V} \sigma dV \;=\; \int\limits_{V_-+V_+} \sigma dV \;+\; \int\limits_{s} \sigma_s dA \qquad\qquad \text{(A.10)}$$

$$\int\limits_{\partial V} \Phi_i n_i dA \;=\; \int\limits_{A_-+A_+} \Phi_i n_i dA + \int\limits_{\partial s} \Phi_i^s n_i ds$$

With these decompositions the equation of balance for the closed system V in Fig. A.2 may be written in the form

$$\int\limits_{V_-+V_+} \frac{\partial \rho\psi}{\partial t} dV + \int\limits_{A_-+A_+} (\rho\psi v_i + \Phi_i)\, n_i dA + \int\limits_{s} [\rho\psi]\, u_i e_i dA \;-\; \int\limits_{V_-+V_+} \sigma dV \;+$$

$$+\frac{d}{dt} \int\limits_{s} \rho_s\psi_s dA + \int\limits_{\partial s} \Phi_i^s n_i ds - \int\limits_{s} \sigma_s dA = 0\,. \qquad\qquad \text{(A.11)}$$

Often the surface-related quantities ψ_s, Φ_i^s and σ_s are zero so that the second line in (A.11) vanishes. For the moment we proceed on that assumption in order to find jump conditions.

We consider the volumes $V_\pm$ to have the shape of a pill box with its flat sides parallel to the singular surface. We let the thickness of that box go to zero first and then the flat sides. In this manner (A.11) reduces to a local jump condition for any point on s

$$[\rho\psi\,(v_i - u_i) + \Phi_i]\, e_i = 0, \qquad\qquad \text{(A.12)}$$

since the fields ρ, ψ, $\mathbf{v}$, $\mathbf{\Phi}$, σ and $\frac{\partial \rho\psi}{\partial t}$ are smooth except on s. As before $[a]$ stands for $a^+ - a^-$ for a generic quantity a.

The jump conditions imply that in each point of the singular surface the influx of Ψ — convective and non-convective — on one side equals the outflux on the other side. This is not always the case and the use of the jump conditions must be judicious. The assumption on the vanishing surface-related quantities becomes precarious in the presence of shear forces in walls, of surface tension and of dissipation in an interface. We have to consider the validity from case to case.

22.5 Examples

The most commonly appearing balance equations of mechanics and thermodynamics are those of mass, momentum, energy, internal energy and entropy. In

Table A.1 Canonical notation for specific values of mass, momentum, energy and entropy and their fluxes and source contributions.

Ψ	ψ	Φ_i	σ		
			σ_{mm}	σ_{mgrav}	σ_{mr}
mass	1	0	0	0	0
momentum	v_j	$-t_{ji}$	0	ρf_j	$\sigma_{mr}^{p_j}$
energy	$u+\frac{1}{2}v^2$	$-t_{ji}v_j+$ $+q_i$	0	$\rho f_j v_j$	$\sigma_{mr}^e=$ $r+\sigma_{mr}^{\pi_j}v_j$
internal energy	u	q_i	$t_{ji}\frac{\partial v_i}{\partial x_j}$	0	$\sigma_{mr}^u=r$
entropy	s	q_i/T	$-\frac{1}{T^2}q_i\frac{\partial T}{\partial x_i}+$ $+\frac{1}{T}t_{ji}\frac{\partial v_i}{\partial x_j}$	0	$\sigma_{mr}^s=\frac{r}{T}$

those cases the generic quantities ψ, Φ_i, σ have concrete physical significance and are all denoted by canonical letters. Table A.1 gives a list.

t_{ji} in the flux density of momentum is called the stress tensor and the flux density q_i of internal energy is called the heat flux. It is customary and appropriate to split the source term σ into three parts of which the first one – denoted by σ_{mm} – is the dissipative source density; that part vanishes for conserved quantities like mass, momentum and energy. Internal energy has such a source, because it may be converted into kinetic energy and vice versa. The entropy has a dissipative source as well and for a viscous, heat-conducting fluid we have calculated its density in Chap. 17; the expression is listed in the table. f_j is the specific body force, usually due to gravitation – here exclusively due to gravitation. As we have seen in Sect. 7.2 the power of that gravitational force may be written as the rate of change of the potential energy. The last column in the table refers to the effects of radiation on matter. Thus $\sigma_{mr}^{p_j}$ is the density of the source of material momentum due to radiation and r is the density of the source of internal energy due to radiation. r/T is the corresponding entropy source density.

We do not always in this book accept the specific forms of the entropic terms in the table. They are the conventionally assumed ones, but in some parts of this book the kinetic theory of gases forces us to take exception, cf. Sect. 3.5 and Chap. 15.

23

References and annotations

Chapter 1 (Clausius)

Carnot's results on the efficiency of heat engines, quoted in Sect. 1.1, appear in his memoir

> S. CARNOT (1824). Réflexious sur la puissance motrice du feu et sur les machines propres à développer cette puissance. A Paris chez Bachelier, libraire Quai des Augustins No.**55**.

In Chap. 21 we have quoted extensively from that memoir in order to show that Carnot indeed expressed the important – and correct – results about maximality and universality of the efficieny of Carnot engines. Also in Chap. 21 we have introduced Carnot as a firm adherent of the caloric theory by quoting a long paragraph from his book. If that theory had been right, no work could have been produced by any heat engine according to the 1st law which, of course, Carnot did not know.

In 1850 Clausius wrote his first paper on thermodynamics.

> R. CLAUSIUS (1850). Über die bewegende Kraft der Wärme und die Gesetze welche sich daraus für die Wärmelehre selbst ableiten lassen. *Poggendorff's Annalen der Physik,* **78**.

Both articles, Carnot's and Clausius' have been translated into English. The English versions are reproduced in a book by Mendoza.

> E. MENDOZA (ed) (1960). Reflections on the motive power of fire by Sadi Carnot and other papers on the Second law of Thermodynamics by É. Clapeyron and R. Clausius. *Dover Publ. New York.*

In 1850 the 1st law was known. Clausius is therefore able to correct Carnot's reasoning in several points. His paper is interesting since it contains the first vestige of the 2nd law about the impossibility of heat to pass from cold to hot.

However, that observation is just mentioned in passing in 1850; its fundamental importance is not recognized until 1854 when Clausius makes it into the salient point of his whole argument. This happens in his paper

> R. CLAUSIUS (1854). Über eine veränderte Form des zweiten Hauptsatzes der mechanischen Wärmethoerie. *Poggendorff's Annalen der Physik* **93**.
>
> Translation: (1856). On a modified form of the second fundamental principle in the mechanical theory of heat. *Philosophical Magazine* (4) **12**.

In that paper the second law is given the form quoted by us in Sect. 1.3 and Clausius obtains the important result $\int \frac{dQ}{T} \leq 0$, equivalent to our formula (1.22). [It is true that his inequality is inverted. But that is only because he counts heat "added" as negative.]

There is no mention of entropy yet. That comes later − in 1864 − in the paper

> R. CLAUSIUS (1864). Über verschiedene für die Anwendung bequeme Formen der Hauptgleichungen der mechanischen Wärmetheorie. *Poggendorff's Annalen der Physik* **125**.
>
> Translation by R.B. Lindsay: (1976). On different forms of the fundamental equations of the mechanical theory of heat and their convenience for application. In: *The Second Law of Thermodynamics*. J. Kestin (ed.), Stroudsburgh (Pa.), Dowden Hutchinson and Ross.

Here Clausius gives the motivation for the choice of the word "entropy" which we have quoted in Sect. 1.3. He also says that he has intentionally chosen the word so as to be similar to "energy" because he feels that the two quantities "are closely related in their physical meaning".

It is also this paper of 1864 which Clausius concludes with the triumphant − and much publicized − statement about the energy and entropy of the universe, which we have cited in Sect. 7.1.

Later in his life Clausius has collected all his papers in a book which contains − often as lengthy footnotes − interesting comments and rebuttals of the numerous challenges which he and his ideas were exposed to.

> R. CLAUSIUS (1887). *Mechanische Wärmetheorie* 1. Bd. (3. Aufl.) Vieweg Braunschweig.

Chapter 2 (Carathéodory)

It is often said that, in order to learn about a subject properly, one should go back to the original literature. This may often be true but it is not so for Carathéodory's papers on thermodynamics. Nobody felt comfortable with them.

C. CARATHÉODORY (1909). Untersuchungen über die Grundlagen der Thermodynamik. *Mathematische Annalen 67.*

C. CARATHÉODORY (1925). Über die Bestimmung der Energie und der absoluten Temperatur mit Hilfe von reversiblen Prozessen. *Sitzgsber. d. preuß. Akad. d. Wissenschaften. Phys.math.Kl.*

Sommerfeld in his textbook dismisses the arguments as "abstract", which to a physicist at that time was a dirty word.

A. SOMMERFELD (1952). *Vorlesungen über theoretische Physik, Bd. V. Thermodynamik und Statistik.* Klemm, Wiesbaden.

Born takes it upon himself to translate Carathéodory's ideas *ad usum physicorum.*

M. BORN (1921). Kritische Betrachtungen zur traditionellen Darstellung der Thermodynamik. *Physik. Zeitschrift* **22**, S. 128, 249, 283.

Born praises Clausius' arguments as "exerting a strong intellectual appeal (because of their) wonderful concepts which lead to the thermodynamic laws in a near magic manner". Then he continues by advertising Carathéodory's procedure because "it is more transparent and uses "normal" mathematics that everybody has learned."

We too have edited Carathéodory's derivation of the 2nd law by restricting attention to homogeneous processes in viscoelastic bodies so that the only dissipative entropy source is internal relaxation like creep. Our argument follows a section of Müller's book

I. MÜLLER (1985). *Thermodynamics.* Interaction of Mechanics and Mathematics Series. Pitman, London, p. 107 ff.

Our main purpose in talking about Carathéodory's work at all was to introduce his crystal-clear argument about the empirical and absolute temperature. That argument was anticipated by Kelvin as early as 1848.

W. THOMSON (later Lord Kelvin) (1848). On an absolute thermometric scale founded on Carnot's theory of the motive power of heat and calculated from Regnault's observations. *Philosophical Magazine* (3) **33**, 313-317.

Kelvin makes use of Carnot's correct statement on the universality of the efficiency of the Carnot engine. His arguments are critically reviewed by Truesdell

C.A. TRUESDELL (1980). The tragicomical history of thermodynamics 1822-1854. Springer, New York, p. 171 ff.

Truesdell makes no attempt to redress Kelvin's reasoning in modern terms; instead he uses it to illustrate the tragicomical aspects of the development of thermodynamics prior to 1854.

Chapter 3 (Boltzmann)

The development of the kinetic theory of gases has been well-documented in two volumes of original texts published by Brush

S.G. BRUSH (1965). *Kinetic Theory Vol. 1 and Vol. 2*. Pergamon Press, Oxford.

We learn from these that Clausius was one of the first contributors to the then new and modern kinetic theory in the 1850's. He is little known for that, because his work was eclipsed and vastly extended by the memoirs of Maxwell

J.C. MAXWELL (1860). Illustrations of the dynamical theory of gases. *Philosophical Magazine* **19**, p. 19-32.

J.C. MAXWELL (1867). On the dynamical theory of gases. *Philosophical Magazine* (4) **45**, p. 129-145.

It was Boltzmann, however, who wrote the integro-differential equation for the distribution function, which we now call the Boltzmann equation – our equation (3.2). Its right hand side represents the "Stoßzahlansatz", a formidable word – even in German – which has defied translation.

L.E. BOLTZMANN (1872). Weitere Studien über das Wärmegleichgewicht unter Gasmolekülen. *Sitzungsberichte Akad. Wiss. Wien* (II) 66.

In this paper Boltzmann identifies the entropy of a gas in terms of the distribution function, our equation (3.20).

That discovery led to the decades-long controversy which we have briefly reviewed in Chap. 21. There is no mention yet in the paper of the probabilistic interpretation of entropy as $S = k \ln W$ with W as the number of possibilities in which a distribution may be realized. That interpretation was probably hammered out in the heat of the controversies that Boltzmann was involved in. We do not know exactly when this happened.

But in 1885 that aspect was there. In his book

L.E. BOLTZMANN (1885). Vorlesungen über Gastheorie I. Teil. Verlag Barth, Leipzig

Boltzmann talks about probabilities of drawing black, white, blue, red etc. balls from an urn and about *a priori* equal probabilities for all combinations. He also points out the analogy of this combinatorial problem to the distribution of molecules in a gas. Therefore Boltzmann knew about $S = k \ln W$ without ever writing that formula, or so it seems.

Modern accounts of the kinetic theory include

S.C. CHAPMAN & T.C. COWLING (1952) 2nd ed. *The Mathematical Theory of Non-Uniform Gases.* Cambridge University Press.

H. GRAD (1958). Principles of the Kinetic Theory of Gases. *Handbuch der Physik,* XII. Springer, Berlin, Heidelbergp. 129-145.

L. WALDMANN (1958). Transporterscheinungen in Gasen von mittlerem Druck. *Handbuch der Physik,* XII. Springer, Berlin, Heidelbergp. 129-145.

C. TRUESDELL, R.G. MUNCASTER (1980). *Fundamentals of Maxwell's Kinetic Theory of a Simple Monatomic Gas.* Academic Press, New York, London.

It remains to discuss the quantization, by which the smallest occupiable element of the $(\mathbf{x}, \mathbf{c})$-space is $\frac{1}{XY}$, cf. (3.30). In Chap. 14 we have related the size of that element to the Planck constant h. And it was indeed Planck who first anticipated the quantization of the $(\mathbf{x}, \mathbf{c})$-space, or the phase space, spanned by $(\mathbf{x}, \mathbf{p})$. In 1911 he gave a lecture to chemists.

M. PLANCK (1912). Über neuere thermodynamische Theorien (Nernst'sches Wärmetheorem und Quantenhypothese). Vortrag in der Deutschen Chemischen Gesellschaft in Berlin. *Akad. Verlagsges. Leipzig.*

In that lecture Planck postulates hypothetical finite "elementary regions" lest the entropy constant be infinite, cf. (3.38). However, says Planck, it seems doubtful or even impossible that such finite regions have any physical significance. We must remember that Planck's lecture was given before de Broglie waves were conceived of. These made sense out of Planck's "elementary regions" as we have shown in Chap. 16.

In Sect. 3.7 and Sect. 16.2 we have discussed the arbitrariness of "smuggling" the term $\ln N!$ into the kinetic equation for S and we have related that term to what we call the pseudo-Gibbs-paradox. The term was thought to be necessary for a comprehensible statistical interpretation of entropy and it was missing in Boltzmann's work. According to Fast

J.D. FAST (1960). *Entropie. Die Bedeutung des Entropiebegriffes und seine Anwendung in Wissenschaft und Technik.* Philips' Technische Bibliothek. Also availabe in Dutch, English, French and Spanish.

the pioneers Planck, Einstein, Ehrenfest and Schrödinger discussed the issue among themselves. Ironically, they thought that the $N!$ term came out of Boltzmann's work and they looked for reasons to omit it. We have explained the true circumstances in Chap. 3 and Chap. 16.

Chapter 4 (Equations of state)

We do not quote anybody specifically for the thermal and caloric equations of state of an ideal gas. These are due to the experimental works of R. Boyle (1627-1691), E. Mariotte (1620-1684), J.L. Gay-Lussac (1778-1850), J.P. Joule (1818-1889) and W. Thomson (Lord Kelvin) (1824-1907). Their findings are now reported in high school textbooks − often in so much detail that the students become quite confused.

For van der Waals' ingeneous arguments, however, we quote his thesis

> J.D. VAN DER WAALS (1873). *Die Kontinuität des gasförmigen und flüssigen Zustandes.* Dissertation, Leiden.

While the van der Waals equation is often written and used, the statistical arguments, that have led to it, are seldom reproduced. One textbook where this is done is

> R. BECKER (1960). *Theorie der Wärme.* Springer, Heidelberg.

Chapter 5 (Gases and Rubber)

One of the authors of this book (I.M.) has recently reviewed the extraordinary properties of rubber in a small book.

> I. MÜLLER, P. STREHLOW (2004). *Rubber and Rubber Balloons − Paradigms of thermodynamics.* Lecture Notes in Physics. Springer, Heidelberg.

In that booklet the literature on rubber and entropy-induced rubber elasticity is reviewed extensively. Here we cite only the three salient papers − all written in the 1930's − which revealed the microscopic composition of rubber and made use of it.

> W. KUHN (1934). Über die Gestalt fadenförmiger Moleküle in Lösungen. *Kolloidzeitschrift* **68**, p.2.

> K.H. MEYER & C. FERRI (1935). Sur l'élasticité du caoutchouc. *Helv. Chim. Acta* **29**, p. 570.

> W. KUHN (1936). Beziehungen zwischen Molekülgröße, statistischer Molekülgestalt und elastischen Eigenschaften hochpolymerer Stoffe. *Kolloidzeitschrift* **76**, p.258.

Modern descriptions of the subject are contained in the monographs by Flory and by Bird, Armstrong & Hassager.

P.J. FLORY (1953). *Principles of Polymer Chemistry.* Cornell University Press, Ithaca, London.

R.B. BIRD, R.C. ARMSTRONG, O. HASSAGER (1976). *Dynamics of Polymeric Liquids. Vol. I Fluid Mechanics, Vol. II Kinetic Theory, J. Wiley & Sons, New York, London.*

Chapter 6 (Statistical thermodynamics)

Statistical mechanics proper — as different from the kinetic theory of gases — goes back to Gibbs and it has led to a rich literature.

J.W. GIBBS (1902). *Elementary principles in statistical mechanics, developed with especial reference to the rational foundation of thermodynamics.* New York and London.

The literature on statistical mechanics and thermodynamics is overabundent and every physics library contains several competently written monographs on the subject. Indeed, the old-fashioned course "Thermodynamics" has all but disappeared from the physics curriculum; it has been replaced by a course called "Thermodynamics and Statistical Mechanics" which is 10% thermodynamics and 90% statistical mechanics.

Ensembles abound in such monographs and few authors take pause to explain why they are needed. A notable exception is Schrödinger in a written account of thoughtful seminars which he gave.

E. SCHRÖDINGER (1948). Statistical thermodynamics. A course of seminar lectures. Cambridge at the University Press.

After having explained that the Boltzmann view of entropy is immediately useful only for systems of independent elements he explains the concept of an ensemble thus: We quote:

> ... Here the ν identical systems are mental copies of the one system under consideration — of the one macroscopic device that is actually erected on our laboratory table. Now what on earth could it mean, physically, to distribute a given amount of energy ε over these ν mental copies? The idea is in my view, that you can, of course, imagine that you really had ν copies of your system, that they really were in "weak interaction" with each other, but isolated from the rest of the world. Fixing your attention on one of them, you find it in a peculiar kind of "heat bath" which consists of the $\nu - 1$ others.

Schrödinger's booklet is well worth reading, because it carefully explains the concepts which other, less thoughtful authors take for granted or ignore entirely.

Chapter 7 through 12 (Entropy vs. Energy)

The chapters 7 through 12 have textbook-character and, if there is anything new in them, it is our manner of presentation which emphasizes the competition of energy and entropy. That is the guiding theme of this book, of course. The idea is not new, although we have never seen it demonstrated in quite the repetitive and systematic form in which we do it − in these chapters − for different examples from elementary thermodynamics and mechanics to physico-chemistry, chemical engineering, and chemistry.

Therefore there is no occasion to cite anyone or any other book in particular. There is one exception, however, and that concerns our treatment of photosynthesis where we follow a paper by Klippel and Müller

> A. KLIPPEL, I. MÜLLER (1997). Plant growth − a thermodynamicist's view. Cont. Mech. & Thermodynamics 9.

This paper exploits the non-orthodox idea that the balance of entropy for the reaction is put right by the entropy of mixing of the evaporated water with the surrounding air. The *orthodox* idea − often cited and never made specific − goes back to Schrödinger

> E. SCHRÖDINGER (1944). What is life? The physical aspect of the living cell. Cambridge at the University Press.

Schrödinger proposes that the necessary entropy increase of the system is balanced by the increase of entropy between incoming and outgoing radiation. He says:

> "Plants... possess their greatest supply of "negative entropy" in the sunlight" and
> "...they drink order so as to avoid the decomposition into atomic chaos."

Those sentences gave rise to one nemesis − one of many − of entropy research, the *negentropy*, much beloved by physicists of the more esoteric type.[1] Actually Schrödinger lived to regret having created that notion. In the German translation of his book in 1948 he says:

> "My remarks about *negative entropy* have been criticized by experts in physics. I have to say to them that I should have used the word *free energy*, if I had spoken to them."

---

[1] Other such notions are: numerical entropies, entropy of hyperbolic systems, entropy and information.

As we have seen in Chaps. 18 and 19 it is indeed true that outgoing radiation has a much larger entropy than incoming radiation. However, it is not clear how the chemical process of photosynthesis could profit from that entropic growth. To our knowledge the first authors to publish doubts about the effect of radiation on the entropy change during photosynthesis were two Japanese authors

A. TSUCHIDA, T. MUROTA (1997). Fundamentals in the entropy theory of the ecocycle and human economy. Environmental Economics.

They assumed that the entropy increase associated with the evaporation of water sets the entropic balance right. This alone, however, is also not good enough because the free enthalpy does not change during the evaporation. To be sure, the entropy grows but so does the enthalpy.

This issue clearly deserved further thought and even now it can hardly be considered closed. Indeed we invite criticism because, after all − as we already indicated in Sect. 12.7 − water plants can hardly be expected to satisfy the entropy balance of photosynthesis by evaporating water!

Chapter 13 (Shape memory)

At this time there are tens of thousands of published papers on shape memory. Metallurgists, physicists and mathematicians have laboured on that field and dozens of international − and interdisciplinary − congresses have been devoted to it. One of the authors of this book (I.M.), and his co-workers, have been contributing actively to the literature on the subject, and Chap. 13 represents mostly a summary of the activities of that group. Therefore we quote their pertinent papers.

M. ACHENBACH (1989). A Model for an alloy with shape memory. *Int. J. of Plasticity* **5**.

Y. HUO, I. MÜLLER (1993). Non-equilibrium thermodynamics of pseudo-elasticity. *Cont. Mech. & Thermodyn.* **5**.

I. MÜLLER, S. SEELECKE (2001). Thermodynamic Aspects of Shape Memory Alloys. *Mathem. and Comp. Modelling* **34**.

O. KASTNER (2003). Molecular dynamics of a 2D model for the shape memory effect. Part I: Model and simulation. *Cont. Mech. & Thermodyn.* **15**.

O. KASTNER (2004). Molecular dynamics of a 2D model for the shape memory effect. Part II: Thermodynamic investigation. *Cont. Mech. & Thermodyn.* (submitted).

Kastner's work is on the molecular dynamics of a 2-dimensional shape memory model and his results may be viewed on the accompanying CD.

Chapter 14 (Third law)

Walter Nernst is usually credited with a forerunner of the third law of thermo-dynamics. And he did state that the entropy of each body becomes independent of pressure as the absolute temperature tends to zero; and that the value which it approaches is independent of the phase of the body. [One must know that many solids may have different phases, i.e. different crystalline structure, at the same pressure and temperature. Some may even be either crystalline or amorphous.] This so-called Nernst theorem – sometimes called 3rd law – was published in 1906.

> W. NERNST (1906) Über die Berechnungen chemischer Gleichgewichte aus thermodynamischen Messungen. *Kgl. Ges. d. Wiss. Göttingen* **1**.

Nernst kept explaining his law – or "theorem" – and he was still at it in his book

> W. NERNST (1924). Die theoretischen und experimentellen Grundlagen des neuen Wärmesatzes (2. Aufl.) Verlag W. Knapp. Halle (Saale).

It is not easy to understand the proof of the theorem, if indeed it can be found in the book. Fortunately, however, we do not have to, because Planck strenghtened the 3rd law and put it into the form which we have quoted in Sect. 14.2. In that form the law could be checked for any given body by integrating specific heats and summing latent heats from the ideal gas state all the way down to $T \approx 0$. It turned out that the third law is not always true. Some substances, like methane, tetrachlormethane, or phosgene have a "zero-point entropy" according to the books where the results of such measurements are reported, e.g.

> E. D'ANS, LAX (1967). Taschenbuch für Chemiker und Physiker, Bd. I. Makroskopische physikalisch-chemische Eigenschaften. Springer Berlin, Heidelberg, New York.

It is assumed that those substances form an amorphous solid phase and that the third law holds only for crystalline solids.

Chapter 15 (Zeroth law of thermodynamics)

Temperature is usually taken for granted and certainly it was taken for granted by the pioneers of thermodynamics throughout the centuries. However, occasionally a bright student asks the question: "Why do we bother with temperature when that quantity does not occur in the basic equations, namely those for mass, momentum and energy?" The answer is that — unlike the densities of mass, momentum and energy — the temperature is measurable. And it is measurable, because it is continuous at a boundary.

When equilibrium thermodynamics was essentially complete as a science in about 1920, the axiomatizers took over. They recognized the fundamental importance of the continuity assumption for temperature. Since, however, the first second and third laws were already firmly labeled, they suggested to call the continuity of temperature the *zeroth law*.

To our knowledge Carathéodory was the first to clearly state the continuity of temperature as its defining property. In a way he expressed what everybody knew anyway. However, his statement was restricted to equilibrium.

We have explained in Chap.15, how the continuity of temperature may be considered as a corollary of the continuity of the entropy flux and its relation to the heat flux. That interpretation provides a new view of temperature which was first conceived by

I. MÜLLER, T. RUGGERI (2004). Stationary heat conduction in radially symmetric situations — an application of extended thermodynamics. *Journal of Non-Newtonian Fluid Mechanics*. Special issue dedicated to the memory of Frank Leslie (R. Atkin, I. Steward (eds.)).

The main consequence of the argument in that paper is that in non-equilibrium the temperature of a gas is no longer a measure for the mean kinetic energy of the atoms of the gas. The reason for the distinction is that the entropy flux in non-equilibrium is no longer universally related to the heat flux by the kinetic temperature according to the kinetic theory of gases.

The above-quoted paper is also remarkable for the fact that it took four years to be published — not because of bad reviews, but "just so". In the meantime Müller and Strehlow had adapted the contents for bringing it to the attention of experimentalists who might wish to establish the difference between the kinetic and thermodynamic temperatures experimentally.

I. MÜLLER, P. STREHLOW (2003). Kinetic Temperature and Thermodynamic Temperature. *Temperature: Its Measurement and Control in Science and Industry*, 7. (Dean C. Ripple (ed.)). American Institute of Physics.

Chapter 16 (Gibbs paradox and degenerate gases)

We have tried to explain that there are *two* Gibbs paradoxa which we have distinguished by calling one of them the pseudo-Gibbs-paradox. It is this one that is most often discussed and which has been resolved in a satisfactory manner. The other one, the true Gibbs paradox is rarely mentioned and it persists as a paradox to this day.

For an instructive discussion of the pseudo-Gibbs-paradox – and its reso-lution – we like to refer the reader to the booklet of Schrödinger which was already cited as a reference pertaining to Chap.6. Schrödinger talks about the new statistics and says:

> "It has always been believed that Gibbs' paradox embodied profound thought. That it was intimately linked up with something so important and entirely new could hardly be foreseen."

Of course he is discussing the pseudo-Gibbs-paradox; he does not give a recipe for avoiding the proper Gibbs paradox; nobody has done that.

The way in which we have developed the new statistics of Bose and Fermi by allowing d atoms to occupy a point $\mathbf{xc}$ is due to Gentile

> G. GENTILE (1940). Osservazioni sopra le statistiche intermedie. *Nuovo Cimento* **17**, p. 493-497.

We believe that only the cases $d = 1$ and $d = \infty$ are realistic possibilities and therefore we have specialized the analysis to those cases after taking advantage of Gentile's elegant method. Gentile himself carries the analysis through to final results for N, U and S_E.

Of course, Gentile also obtains our formula (16.33) as the condition for the classical limit. Once again it was Schrödinger who discussed that formula in the most instructive manner. He calls it

(a) satisfactory, (b) disappointing, (c) astonishing.

The result is *satisfactory*, because it guarantees that for high tempera-ture and small densities we obtain well-established experimental results. It is *disappointing* because the condition implies that the effects of degeneration in a gas should occur in a range of density and temperature where van der Waals forces have long led to condensation. And the result is *astonishing*, be-cause it requires scarce occupation of the element $d\mathbf{x}d\mathbf{c}$, so that factorials for $N_{\mathbf{xc}} \ln N_{\mathbf{xc}} - N_{\mathbf{xc}}$ are out, while classical statistical mechanics is firmly based on factorials.

Chapter 17 (Thermodynamics of irreversible processes)

Sofar we have never specified what makes entropy grow and free energy de-crease. To be sure we have loosely stated that heat conduction and internal friction are responsible, cf. Sec. 1.3, but those remarks were unspecific. It takes thermodynamics of irreversible processes to provide formulae for the entropy source.

Thermodynamics of irreversible processes converts thermodynamics into a field theory for the determination of the fields mass density $\rho(\mathbf{x})$, velocity $v_i(\mathbf{x}, t)$ and temperature $T(\mathbf{x}, t)$. Early versions of such a field theory were presented by Jaumann and Lohr.

G. JAUMANN (1911). Geschlossenes System physikalischer und chemischer Differentialgesetze. *Sitzungsbericht Akad. Wiss. Wien,* **12** (IIa).

E. LOHR (1926). Entropie und geschlossenes Gleichungssystem. *Denkschrift Akad. Wiss. Wien,* **93**.

These memoirs did not have much impact so that we may say that Thermodynamics of Irreversible Processes really started with three papers by Eckart.

C. ECKART (1940). The thermodynamics of irreversible processes I: The simple fluid. *Phys. Rev.* **58**.

C. ECKART (1940). The thermodynamics of irreversible processes II: Fluid mixtures. *Phys. Rev.* **58**.

C. ECKART (1940). The thermodynamics of irreversible processes III: Relativistic theory of the simple fluid. *Phys. Rev.* **58**.

Chap. 17 gives a brief account of the first of those papers. In the 1960's non-equilibrium thermodynamics experienced a renaissance based on a paper by

B.D. COLEMAN, W. NOLL (1963). The Thermodynamics of Elastic Materials with Heat Conduction and Viscosity. *Arch. Rational Mech. Anal.* **13**.

One of the authors of the present book (I.M.) contributed some ideas to this new version of thermodynamics which was reviewed in 1984 by Truesdell

C. TRUESDELL (1984). *Rational Thermodynamics* (2nd ed.). Springer New York, Heidelberg.

Concerning the wretched principle of minimum entropy production, this is due to Prigogine and it is described − as much as it can be described − in the book by

P. GLANSDORFF, PRIGOGINE, I. (1971). *Thermodynamic Theory of Structure, Stability and Fluctuations.* Wiley Interscience, London.

The explicit proof that the principle contradicts the 1st law of thermodynamics was given by Barbera. She treated heat conduction and shearing motion.

E. BARBERA (1999). On the principle of minimum entropy production for Navier-Stokes-Fourier fluids. *Cont. Mech. & Thermodyn.* **11**.

Chapter 18 and 19 (Radiation Thermodynamics)

Once the photon model of radiation was conceived it has provided a convenient manner of deriving the balance laws for radiative transfer from the kinetic theory of photons. We have learned the theory from

S. CHANDRASEKHAR (1960). *Radiative Transfer.* Dover Publ. London.

and we have reproduced it here in a compact form.

We are mostly interested in the entropic radiative terms and we do not wish to enter into a discussion of the details of absorption, emission and dispersion of radiation. Therefore we restrict the attention to stationary processes, where the entropy source may be calculated from the balance of the in- and outgoing entropy fluxes which are known, or have been measured or estimated.

In the calculation of the entropy source of the earth and its decomposition we follow the arguments put forward by Weiss

W. WEISS (1996). The balance of entropy on earth. *Cont. Mech. & Thermodyn.* **8**.

One result emphasized by Weiss — and also in this book — is that the dissipative entropy source of matter is a lot smaller than the entropy source of radiation due to its interaction with matter. Weiss considers the former part — the dissipative entropy source — as the relevant one for meteorology, life science and ecology. The distinction between the two parts is often not made in the literature on radiation thermodynamics. See for instance

W. EBELING (1991). Chaos — Ordnung — Information. Harri Deutsch, Thun-Frankfurt/Main.

A. STAHL (1993). Entropy and Environment. In: *Statistical Physics and Thermodynamics of Nonlinear Nonequilibrium Systems.* (W. Ebeling, W. Muschik, eds.) World Scientific Publ. Co.

The fluxes which we use to calculate the entropy source of the earth and its atmosphere have been taken from the meteorological literature.

M. KLEEMANN, M. MELISS (1988). *Regenerative Energiequellen.* Springer Berlin, Heidelberg.

We have been interested in the possibility of defining a radiation temperature Ξ by relating the entropy source of radiation σ^s_{rm} to the radiative source r of internal energy through the simple equation $\sigma^s_{rm} = -\frac{r}{\Xi}$. In this book this is done for thin layers of matter of different temperatures and in radiative contact. A more general treatment of this idea by Müller may be found in the book

I. MÜLLER, T. RUGGERI (1998). *Rational Extended Thermodynamics.* (2nd ed) Springer, New York.

Chapter 20 (Sociobiology)

A life-long preoccupation with thermodynamics has given us the fixed idea that thermodynamic ideas and thermodynamic arguments may be applied outside physics — in sociology and economy. We are not the only ones with such an idea, nor the first. Indeed, Mimkes has already conceived of miscibility gaps in society

J. MIMKES (1995). Binary Alloys as a Model for the Multicultural Society. *J. of Thermal Analysis.* **43**.

What is new in our case is the connection to game theory which furnishes the possibility to relate the occurrence of the miscibility gap to the behavioural peculiarities of the species rather than just conceive its existence.

It seems that the hawk-dove game was invented by Maynard-Smith and Price as a model for an evolutionary strategy allowing for a stable mixed population

J. MAYNARD-SMITH, G.R. PRICE (1973). The Logic of Animal Conflict. *Nature.* **246**.

The game is often cited in the literature on socio-biology or just as an interesting paradigm of game theory.

R. DAWKINS (1989). *The Selfish Gene.* Oxford University Press.

P.D. STRAFFIN (1993). *Game Theory and Strategy.* New Mathematical Library. The Math. Assoc. of America, **36**.

The present extrapolation of the game to two strategies has been proposed by Müller and he has also embedded the game firmly into thermodynamics with all its ingredients, like the first and second laws.

I. MÜLLER (2002). Socio-thermodynamics — integration and segregation in a population. *Cont. Mech. & Thermodyn.* **14**.

The present treatment is an improved version of that work, because it provides more plausible contest strategies.

The person who made entropy popular for a while among economists was Georgescu-Roegen with his book

N. GEORGESCU-ROEGEN (1971). *The Entropy Law and the Economic Process.* Harvard University Press, Cambridge, Mass.

The author puts his profound scientific erudition in evidence and his citations run the gamut from classical Greek philosophers to modern scientists. The study of the book is a highly enjoyable experience except when it comes to the discussion of entropy. Indeed the introduction of entropy is limited to a textbook chapter on Carnot processes and to a weak account of the entropy of mixing. It seems that the author is much impressed with the efficiency of heat engines as an economic criterion.

However, such occasional superficiality does not detract from the joy of reading. Thus, when the author laments the arbitrariness with which the word entropy is assigned to anything and everything:

> "The code of Humpty Dumpty − which allows one to use a word with any meaning one wishes − is much too often invoked as a supreme authority on terminological prerogative. But nobody seems to have protested that ordinarily the only consequence of this prerogative is confusion".

Reading this we cannot help thinking of the "entropy" of hyperbolic systems of differential equations, where entropy is a convex (!) function, or of the "entropy" of numerical analysts which is something or other connected with their schemes. We do protest!

Chapter 21 (History of Thermodynamics)

Most of the information about the emergence of the concept of temperature and its measurement is taken from a book by Knowles-Middleton

W.E. KNOWLES-MIDDLETON (1966). *A History of the Thermometer and its Use in Meteorology* . Johns Hopkins Press. Baltimore .

This excludes the instructive letter written by Giovanfrancesco Sagredo to Galilei. That letter is excerpted from

Le Opere de Galilei Galileo (1902). Ed. Naz. Vol. XII. Tipografia di G. Barbera.

The biographical notes and indiscreet remarks are often taken from Asimov's Biographical Encyclopedia

I. ASIMOV (1975). *Asimov's Biographical Encyclopedia of Science and Technology.* Pan Books, London.

Most of the protraits and some of the appended comments are from Lenard

P. LENARD (1941). *Große Naturforscher.* J.F. Lehmann Verlag München.

Robert Mayer is usually forgotten as one of the founders of the 1st law, but not in his hometown Heilbronn. Their archive has published a biographical book from which we have learned much about Mayer's adversities

H. SCHULZ, H. WECKBACH (1964). *Robert Mayer. Sein Leben und sein Werk.* Veröffentlichung des Archivs der Stadt Heilbronn.

Helmholtz' pamphlet was republished — both in facsimile handwriting and as transcript — only 20 years ago

C. KIRSTEN, K. ZEISLER (Hrg) (1982). *Hermann Helmholtz. Über die Erhaltung der Kraft.* Dokumente der Wissenschaftsgeschichte. Akademie Verlag. Berlin.

The history of thermodynamics after 1854 is an exciting period of natural philosophy in which some of the greatest discoveries in physics were made. As such the period is often visisted — and revisited — by historians of science and the papers of the pioneers are republished and extensively commented. We have quoted the books by Brush in Chap. 3. These books and also Brush's brochure

S.G. BRUSH (1978). *The Temperature of History. Phases of Science and Culture in the Ninetheenth Century.* Burt Franklin, New York

have given us much information which is reflected in the chapter on history.

Index

Printing: Krips bv, Meppel
Binding: Stürtz, Würzburg